Air Pollution, Human Health, and Public Policy

Air Pollution, Human Health, and Public Policy

Charles T. Stewart, Jr.
The George Washington University

Lexington Books
D.C. Heath and Company
Lexington, Massachusetts
Toronto

Library of Congress Cataloging in Publication Data

Stewart, Charles T. 1922-
Air pollution, human health, and public policy.

Includes bibliographical references and index.
1. Air–Pollution–United States. 2. Environmental policy–United States. I. Title. [DNLM: 1. Air pollution–United States. 2. Public policy–United States. WA754 S849a]
TD883.2.S76 614.7'1'0973 78-13818
ISBN 0-669-02670-0

Published simultaneously in Canada

Printed in the United States of America

International Standard Book Number: 0-669-02670-0

Library of Congress Catalog Card Number: 78-13818

Contents

List of Tables

Preface

This inquiry was initiated with an open mind, but with a suspicion that producers short on social responsibility were finally being forced to give due consideration to the consequences of their activities. To conclude, however, the lack of responsible behavior appears more pervasive, extending to some protagonists of environmental purity who seem careless about evidence, extreme in their goals, and indifferent over the costs and consequences of their quest to others.

What is striking about the rather recent emergence of an environmental concern are the widely-held assumptions that underlie it and the attitudes it reflects. There is first the confidence in the power of man to control his environment—that megalomania of an industrial-scientific civilization, faintly reminiscent of the legendary King Canute. Air quality standards will be attained (although a *Wall Street Journal* editorial suggests they may require installing stack scrubbers on volcanoes). Second, there is a touching faith in the power of the law, in this case to create appropriate technology by fiat, to achieve on schedule whatever standards have been decided upon by a democratic process. Motor vehicle emissions *will* be cut by ninety percent. Third, there is the nostalgic view of nature as immaculate until sullied by the hand of man and the growth of industry (although smog was reported in southern California when only swallows came to Capistrano). Fourth, there is the attribution of guilt to those who build and maintain an urban-industrial society, as though those of us who participate or only enjoy it are innocent victims. Finally, there is the failsafe attitude: "do no harm," forgetful that life is a hazardous occupation, that the effort to secure man from all harm has consequences in reduced resources for other purposes that could outweigh the gain from cleaner air, and that this nation was not built nor made secure by risk aversion. This set of attitudes seems characteristic of America in the late twentieth century with environmental anxiety only one of many manifestations of a cultural crisis of confidence.

Nevertheless the problem is real enough, not just a conjuncture of attitudes, but what Fred Hirsch has labelled the "social limits to growth." If pure air is not the end-all, neither is economic growth. Any society with multiple and conflicting goals must learn how to count the costs as well as the benefits of alternative courses, and not allow rhetoric to rule arithmetic. When policies are poorly informed and less than wise, flexibility is a virtue. Environmental issues will be with us a long time, and there is no solution that does not create new problems.

Acknowledgments

I am grateful to Thomas F. Johnson for his encouragement, to Mary Weafer and Corazon Siddayao for research, and to Glenda Howell for guiding the manuscript through the typewriter.

1 Introduction

We seem to be pouring into the environmental quality bottle all our individual and social yearnings for peace, stability, and quiet; for social justice in the world; and for more meaningful lives. To these we have added our passions for reform of values and improved quality of life generally; and our antagonism toward modern industrial growth and abuses by private enterprise. . . . Environmental quality is only a lesser element in these problems, not the whole of them.[1]

Fulminations against air pollution could be heard early in the nineteenth century, directed against the allegedly poisonous fumes of factories and then-new locomotives belching black smoke along the pristine countryside. Thomas Carlyle and John Ruskin were the Ralph Naders and Ezra Mishans of their day. But more scientific evidence of possible damage to health and associated widespread concern are of recent origin. National legislation dealing with air pollution (as distinct from industrial safety legislation affecting particular occupations and industries) is a product of the 1960s, culminating in the Clean Air Act Amendments of 1970.

National ambient air quality standards have been established that may require draconian measures in some urban areas, including severe restrictions on the use of cars and temporary cutbacks in industrial production and generation of electricity. Emission control requirements have been established for a large number of industries, imposing billions of dollars in additional costs on industry and on consumers. Environmental Protection Agency (EPA) policies are affecting the location of industry and regional prospects for economic growth. At issue in the courts is the right of areas with pollution far below admissible levels to worsen their air quality as a by-product of industrial location and economic growth.

Yet standards were established on the basis of uncertain and contradictory evidence. Many instruments relied upon for pollution readings were inaccurate; the monitoring system was inadequate. The resulting measures of air pollution were not measures of exposure to pollution. Studies of excess illness and mortality associated with air pollution report inconsistent findings.

Air pollution is not new, nor has it been getting worse. But many more people are concentrated in large cities today than ever before. Even if cities have no more air pollution than in the past, the exposure to air pollution has increased greatly.

The number of air pollutants is large and growing. Some, such as sulfur and nitrogen oxides and carbon monoxide, are widely distributed, exposing much of the population to their effects. Others, such as dusts causing silicosis among quarry workers and black lung among coal miners, affect small and clearly identifiable groups. Pollution as a by-product of industrial activities is constantly generating new threats to human health, as new chemicals whose consequences are unknown are produced. Most of them are occupational hazards that do not endanger the general population.

This report is concerned only with the more obiquitous sources of air pollution—some of recent origin and only recently the subjects of national legislation. In the case of ambient air pollution, it is difficult to establish damage to health because of lower concentrations; it is difficult to identify the individuals who have been or will be harmed; and it is sometimes impossible and rarely easy to establish a causal connection between air pollution and health loss. Furthermore, identification of the sources of pollution may prove difficult, and inspection and control are large jobs, considering the many kinds and sources of air pollution. The air is not a completely separate domain. Some harmful air pollutants are eventually deposited on the earth as acid rain, which may be a contributory factor to high rates of death from heart disease in the Northeast.

The essence of the problem of ambient air pollution is that the harmful effects or risks are widely dispersed, whereas benefits of polluting industry (more local jobs, lower production costs) may be narrowly concentrated. The beneficiary bears little if any of the cost. The victims cannot be readily identified, or the harm done to them readily quantified. Under these conditions, the private citizen or the private firm acting alone is unable to solve the problem, and neither has much incentive to do anything about it. Thus public action—standards, incentives, and deterrents—becomes necessary once the public is alerted to and concerned with the problem.

A direct challenge to clean air goals and the methods of attaining them was presented by the energy crisis that developed in 1973, the year of the Arab oil embargo, followed by a quadrupling of the price of OPEC oil. The implications became apparent only slowly, as the federal government groped toward an energy policy that lays primary stress on substitution of coal for oil. The energy crisis has escalated the costs of clean air. The goals of clean air and energy self-sufficiency cannot be reconciled. They are too disparate to be subject to a common calculus. The outcome of their political confrontation is uncertain at this time.

This report assesses the factual basis for action and evaluates public policy in the light of the available evidence. Chapter 2 distinguishes between pollutant emissions, air pollution, and exposure to air pollution. It reviews our capability to measure air pollution, today and at the time standards were set. Chapter 3 surveys the literature on the effects of widespread air pollutants on health, and on the diseases associated with air pollution. Chapter 4 considers the basis for setting national emission standards and national ambient air quality standards, and the current status of federal regulation of air pollution. Chapters 5 and 6 discuss the difficulties of quantifying the benefits and the costs of clean air legislation, and present some estimates that leave the issue in doubt. Chapter 7 considers possible consequences of clean air policies for the growth prospects of areas with clean air and with polluted air. Chapter 8 addresses the conflict between clean air and energy policies. Finally, Chapter 9 considers the relative merits of regulation and alternative approaches via price, tax, and subsidy policies to achieve acceptable air quality.

2 Emission, Pollution, and Exposure

Major Air Pollutants and Their Sources

The primary air pollutants found in most urban areas are carbon monoxide, nitrogen oxides, sulfur oxides, hydrocarbons, and particulate matter (both solid and liquid). Table 2-1 presents estimated emissions by major source. The two sources ubiquitous in urban areas are transportation (predominantly automobiles) and fuel combustion in stationary sources, including residential, commercial, and industrial heating and cooling and thermoelectric power plants. Motor vehicles are the main source of man-made carbon monoxide (CO) and a major source of hydrocarbons (HC) and nitrogen oxides (NO_X). Fuel combustion in stationary sources, on the other hand, is the dominant source of sulfur dioxide (SO_2) and a major source of nitrogen oxides and particulates (see table 2-1).

Industrial processes, the main source of man-made particulates and an important source of carbon monoxide, hydrocarbons, and sulfur dioxide, are less uniformly distributed among urban areas than either motor vehicles or stationary sources. The primary metals industry accounts for a large proportion of particulates and sulfur oxides emitted by industry. It is regionally concentrated, so that for most areas it does not pose a pollution problem. Furthermore, the emission sources are few in number compared to motor vehicles or stationary sources.

Different types of fuel produce the above-mentioned and other pollutants in quite different proportions as well as in different amounts per Btu. Gas is cleanest, whereas coal is especially dirty in terms of particulates and sulfur. Coal and residual oil and other petroleum-based fuels vary widely in sulfur content. They are all substitute fuels in stationary sources. The energy crisis, resulting in pressure to use higher-sulfur oil and to switch from oil to coal, has serious implications for the level and composition of air pollution. Gasoline additives are a source of airborne lead, which is not associated with other fuels.

The volume of particulates emitted is not a meaningful indicator of pollution. What matters is the chemical composition of particulates, some of which appear harmless, others quite harmful. Size is also important; the larger the size of a particle, the shorter the distance it will travel.[1] Thus, a given weight of particulate emissions in small sizes will result in a much higher level of air pollution over a larger area than the same weight in large particles. A major portion of particulates with diameters of less than 0.5 micrometers is deposited

Table 2–1
Estimated Man-made Emissions of Air Pollutants in the United States, 1974
(million tons per year)

Source	*Carbon Monoxide*	*Particulates*	*Sulfur Dioxide*	*Hydrocarbons*	*Nitrogen Oxides*
Transportation	73.5	1.3	0.8	12.8	10.7
Fuel combustion in stationary sources	0.9	5.9	24.3	1.7	11.0
Industrial processes	12.7	11.0	6.2	3.1	0.6
Solid waste disposal	2.4	0.5	0.0	0.6	0.1
Miscellaneous	5.1	0.8	0.1	12.2	0.1
Total	94.6	19.5	31.4	30.4	22.5

Source: Environmental Protection Agency, reproduced in *Environmental Quality,* Sixth Annual Report of the Council on Environmental Quality (Washington: U.S. Government Printing Office, 1975), p. 440.

Note: This report includes estimated emissions for 1971 that are substantially different, especially for hydrocarbons, from the figures published in the Fourth Annual Report (p. 266).

in the pulmonary portion of the respiratory system.[2] Particulates from fuel combustion in stationary sources and motor vehicles are primarily carbon or soot. Industrial processes, however, are the main source of particulates, whose composition differs from one industry to another and therefore from place to place.

Hydrocarbons, like particulates, vary widely in composition. Since, however, their largest single origin is the ubiquitous motor vehicle, geographical differences in level and composition of emissions are smaller. Hydrocarbons are also a source of secondary air pollutants resulting from complex chemical and photochemical processes in the atmosphere. In the presence of sunlight, harmless or relatively harmless hydrocarbons react with nitrogen oxides to form photochemical oxidants, of which the most important is ozone. The amount of oxidants in the air cannot be related directly to hydrocarbon or nitrogen oxide emissions, since their formation is a complex, poorly understood process, affected by critical ratios of HC, NO_X, and O_3, by the composition of the hydrocarbons, by sunlight and other weather factors, and by time. Aromatic hydrocarbons, for instance, produce substantially more photochemical aerosol than olefins and paraffins, all of which are found in auto exhaust. "To formulate cost-effective oxidant and NO_X abatement strategies, some understanding is required of the processes and factors that influence the conversion and transport processes and the interaction of NO_X with atmospheric oxidant/HC. The present state of knowledge is very incomplete and urgently needs improvement. . . ."[3]

Two harmful pollutants, lead and asbestos, which were not included among those for which ambient air quality standards were originally established, have been found to be more prevalent in possibly harmful concentrations than was initially suspected. Airborne lead is believed to derive predominantly from the combustion of leaded gasoline. The principal sources of asbestos in urban air are still being evaluated. Among contributors are brake linings, tile, insulation and fireproofing materials, serpentine rock, and other fibrous silicate minerals used in road building and construction. Both lead and asbestos are among the pollutants recommended for worldwide observation as part of EARTHWATCH, the United Nations Environment Program monitoring system.

Natural Versus Man-made Air Pollution

Emission of man-made pollutants does not account for all pollution, but does add to the burden of pollution generated by natural processes. Human production of carbon monoxide is less than one tenth that of natural sources.[4] Man contributes about one third of the total atmospheric emissions of sulfur compounds, but nearly all of the sulfur dioxide.[5] In industrial areas, man-made emissions preponderate. The share of man-made sulfur compounds in air cannot be measured with precision because we lack the means to distinguish man-made from biogenic origins of airborne sulfur compounds, and we lack knowledge of the rates of conversion of H_2S and SO_2 into sulfate particles in polluted as well as unpolluted air. Natural processes account for more than ten times the amount of nitrogen oxides harmful to health—NO and NO_2—produced by man's activities.[6]

Although nature also greatly outproduces man in airborne hydrocarbons, which create the haze in forested areas that gives the Smoky Mountains their name, human production accounts for a large share of the more dangerous compounds, such as benzene.[7] Natural processes of volcanic eruptions and forest fires account for much of the particulate matter in the atmosphere, together with the winds that drive dust from the land and salt from the sea.[8] Man's deforestation and short-sighted agricultural practices have contributed to nature's burden. With particulates as with hydrocarbons, however, man's role is more significant in generating the more harmful substances, such as soot, lead, and asbestos.

The human body does not care what the sources of pollutants are. But knowledge about origins is essential to determine how much can be done about them and how best to do it. The point of making the distinction between natural and man-made pollution is that little can be done to reduce the natural sources of pollution, whereas much, perhaps, can be done to reduce the sources deriving from human activities. Only if (or, more pertinent, where) sources of human origin are dominant can antipollution measures prove effective. This does not rule out all countermeasures, but they must be aimed toward reduction of exposure to air pollutants rather than reduction of pollution, or toward treatment to minimize the harmful effects of exposure.

The sand that blows into people's lungs in arid areas and that damages their corneas is of natural origin. It cannot be regulated or legislated against. However, human activities may contribute by denuding vegetation, as in the Sahel today and the Middle East in historical times, and these acitivites can be regulated. It may be possible to change nature to some degree, but the time scales involved and the scope and nature of the policies and programs are utterly different from those involved in reducing emissions from power plants. Solar proton storms that greatly reduce stratospheric ozone and thereby increase the incidence of skin cancer are quite beyond human control, but the same effects produced by high-flying aircraft are not.

Is pollution airborne or earthbound or water-based? It is widely believed, for instance, that lead absorption in urban areas occurs predominantly through the lungs via the combustion of leaded gasoline. But this only accounts for one third of the lead absorbed by the body; two thirds are ingested with food and water. Formerly, people believed that lead was absorbed largely from lead oxide pigment in paint. Around the turn of the century, when lead concentration in human tissue was higher than it is today, a major source was water coursing through lead pipes, rain collected via lead-lined catchments on roofs, water stored in lead-lined containers.[9]

Tests of Egyptian mummies have revealed low lead concentrations, but mercury concentrations comparable to those found in modern man.[10] High mercury concentrations discovered in swordfish some years ago were widely attributed—mistakenly, it seems—to human industrial and agricultural pollution of the oceans. Tests of large fish caught between 1871 and 1971 indicated no particular trend in mercury concentration, suggesting that mercury in the seas is predominantly of natural origin: the erosion of rock and soil containing mercury.[11] If so, the appropriate policy is not industrial pollution control, but a shift in diet away from large fish high on the food chain, perhaps to plankton, made palatable by being soaked in soy sauce. (Some mercury, however, used as a fungicide in paints and other products, is inhaled.)

Lung cancer is believed to be highly correlated with tar and other hydrocarbon products of combustion. But a recent study suggests that the culprit is radioactive lead deposited on tobacco leaves and concentrated in the lungs.[12] If so, low-tar cigarettes may do little to diminish risk.

Pollutant Emissions Versus Air Pollution

If man-made urban air pollution is often many times higher than natural levels, as it is, the cause is not the amount of pollutants emitted by man, but their concentration in space and the time it takes for natural processes of rain, wind, and chemical change to disperse, dilute, or eliminate them from the urban air. If much air pollution is of natural origin, increases or decreases in man-made air pollutants must have much less than proportional effects on ambient air pollution.

Pollution is not localized at the point of emission. Monitoring of suspended particulate matter in the United Kingdom determined that the most important factor was not the seasonal heating cycle, but wind direction, whether from Europe or from the Atlantic. The smaller particles, which can be carried long distances by wind and which have predominated in observed variations, are sulfate, nitrate, ammonium, and fluoride components. It is estimated that more than two thirds of the sulfur in the air in Sweden is anthropogenic, and that 77 percent of that amount originates outside Sweden.[13] Photochemical pollution resulting from emissions in New York City is transported 300 kilometers, as far as northeastern Massachusetts, by prevailing winds.[14]

Precipitation in the northwestern United States is substantially more acidic than elsewhere in the nation. Sulfur content has declined as a result of the shift from coal to natural gas. However, the NO_X content has increased (NO_X is produced in the combustion of natural gas and motor fuels). The increasing pre valence of taller smokestacks, with precipitators removing the larger particles, is dispersing these pollutants much greater distances than in the past, to hundreds of kilometers from industrial centers.[15]

Whereas the relation between emissions and pollution in any particular locality may depend largely on wind velocity and the presence or absence of temperature inversions, the general effect of emissions on atmospheric pollution (and the local effect, in the absence of wind) depends on the average life of pollutants in the atmosphere: the rate at which they are absorbed by earth and water, precipitated by rainfall or gravity, or altered chemically in the air into harmless products. Our knowledge of the average length of residence of pollutants in the atmosphere is undergoing revision. The mean residence of sulfur has been estimated at two to four days; that of nitrogen oxides, at five days[16]; and of carbon monoxide, at one month.[17] The most common hydrocarbon, methane, may remain in the air for years, although others are absorbed much more quickly. The mean residence of particulates is a meaningless figure, since the range is from minutes to years, depending on composition and size.

The soil, as well as the seas, has an enormous capacity to absorb many air pollutants, including CO, SO_2, NO_X, and hydrocarbons.[18] That is one explanation why man-made emissions, even when they dwarf nature's output, do not appreciably increase air pollutant concentrations except locally and for a short time.

The effects of air pollution on health are not measurable directly in terms of the emission of pollutants. Wind speed and direction, temperature inversions, topography, humidity, and rainfall influence the concentration of pollutants and their local persistence. Climatic conditions also affect the rate at which primary pollutants undergo complex chemical changes in the atmosphere, creating secondary polluting agents, which may be the major sources of damage to health.

Measurement of Exposure to Air Pollution

Most discussions of the effect of air pollution on health assume that the cause is adequately measured; if it is not, then it is not possible for its effect to be adequately measured. The fact is that measurement of air pollution is grossly inadequate for relating pollution to health.

First, some measurements relate to emissions, not to air pollution. The two are connected, but not closely. Second, measuring instruments and techniques are often inaccurate and unreliable. Furthermore, the number of air quality monitoring instruments is often too small to provide an adequate sampling of the area that they serve; the number of observations over time in any particular monitoring instrument may also be too small for adequate sampling of that area.

Fourth, even for areas where these limitations are minimized, reliable time series of air pollution are not available for any pollutant. Hence, all that measurements can confidently indicate is the short-run changes in health that may be associated with measured air pollution. The effects on the probability of cancer would require time series of many years, since the health effect often lags twenty or more years after the cause. Much of the harmful effects in the form of respiratory and cardiovascular ailments may also require years, if not decades, of pollution records.

Fifth, even accurate, detailed measurement of air pollution over extensive time periods does not measure air pollution exposure. To the extent that pollution varies over a metropolitan area, for instance, the exposure of an individual depends on where he lives, works, and spends his leisure time. Furthermore, all measures refer to ambient air pollution. But most people spend much of their lives indoors, where the level and composition of pollution can be quite different. Timothy Crocker, M.D., chairman of the department of community and environmental medicine at the University of California (Irvine), testified:

> Human exposure studies must be acknowledged as needed experimental information. Single pollutants alone cannot be studied. Instead, multiple pollutants at naturally occurring levels must be used. With these, the stresses which coincide with elevated O_3, NO_x, *et al.*, must be combined to re-create the real-life exposure; these stresses include heat, exercise and/or tasks requiring detailed performance, vigilance, perception.[19]

Finally, in considering exposure over extended time periods, account must be taken of the fact that individuals travel between areas with quite different ambient air pollution. Exposure is a characteristic of individuals rather than of areas.

Measuring Instruments

Most air pollution measuring instruments can only sample the air in their immediate vicinity. This limitation results in the need for large numbers of instruments if reliable, statistically meaningful observations for an entire urban area are to be obtained. The situation is more serious for particulates, which may be quickly deposited, than for gases, which may spread relatively evenly throughout an area. But it is serious even for gases whose origin is concentrated.

A second limitation of many measuring instruments is the time required for a single observation. This limits the number of observations over the day and week, and again requires a larger number of instruments ro provide sufficiently frequent observations at all points where they are needed.

There are further limitations in the measuring techniques themselves, apart from the question of adequate sampling over space and time. The most widely used instruments rely on wet chemical techniques, which have high maintenance requirements and lack sensitivity. The main problem is that different reagents are needed for measuring each pollutant, but pollutants other than the one being measured interfere with the chemical reactions so that it is difficult to determine exactly what is being measured.

It was recently confirmed that prevailing chemical techniques for measuring nitrogen dioxide were highly sensitive to the concentration of nitric oxide (NO).[20] As a result, readings were not linearly related to the NO_2 concentration, but progressively overstated the higher concentrations. Consequently, the EPA reconsidered its NO_2 standards. It concluded that of forty-seven areas originally believed to have a NO_x problem serious enough to adopt emission controls for stationary sources, only three had such a problem. It recommended relaxation and postponement of auto emission standards on NO_2.[21]

The trend is to move away from chemical toward physical means of detecting and measuring pollutants, in particular to spectographic techniques.[22] Although such optical techniques are capable of continuous monitoring and are more sensitive for most pollutants than chemical methods, they retain certain disadvantages. Most of them are designed to detect a single pollutant, so that many instruments are still required. Like traditional techniques, they are subject to interference from other pollutants and from water vapor. Instruments that are flexible enough to detect a wide range of pollutants and are sensitive and reliable exist, but they are just out of the development stage and their cost is high.

Very much still in the research and development stage are laser-based instruments that overcome the limitation of nearly all other instruments in measuring pollutants in their immediate vicinity. The tunable laser is capable of measuring pollution at a distance along a straight path. Instruments with an effective range of one to six kilometers have been designed, and others with many times that range, usable from aircraft and eventually from space satellites for observation and measurement of emission sources, have been proposed.[23]

Two methods of measurement for practiculates are currently in use: gravimetric determination of "total suspended matter," and photoelectric assessment of the "black suspended matter."[24] The former is not capable of continuous measurement of particulate weight with any reliability. The minimum period for reliable measurements depends on the volume of the sample and the particulate concentration. Photoelectric assessment, on the other hand, is affected by the degree of darkness of the suspended pollutants. Neither method will provide information on the chemical composition or the size distribution of particulates. Additional techniques (with their own measurement errors, as indicated above) are needed to determine chemical composition.

In sum, most measurements of air pollution to date refer only to the immediate vicinity of the instrument. Even then, their reliability is low because of lack of sensitivity, specificity, and reliability of the instruments themselves.

Monitoring

Air quality monitoring was initiated by the Public Health Service in 1953 for a limited range of pollutants, primarily particulates and airborne proteins (pollens), in seventeen communities. By 1970 there were over 7,000 sampling stations with 14,000 samplers (most of them operated by state and local pollution control agencies), and the number of pollutants monitored had greatly expanded. However, the number of measuring devices for most of the pollutants (except sulfur and particulates) has remained very small.[25] According to Russell Train, chairman of the Council on Environmental Quality, writing in October 1972, "Much of our air pollution data . . . is based on results from only one monitoring station in each community."[26]

It is widely assumed that levels of pollution are much higher near the center of large urban areas than in the suburbs. (Carbon monoxide readings can be as much as five times as high in the central city). Readings in Washington, D.C., however, revealed higher pollutant concentrations in the suburbs.[27] While auto traffic may be much heavier in central cities, factories have long since dispersed to the suburbs. Thus, the composition of air pollution is likely to be different in different parts of the city. This is especially true of particulates, which may travel only short distances before falling to earth.[28] More detailed knowledge of the distribution of pollutants throughout metropolitan areas is needed than can be provided by a single monitoring station, or by station sites not representative of the distribution of population. Otherwise, we really do not know even approximately the exposure of the population to air pollution. However, many monitoring instruments are poorly located as well as poorly maintained.[29]

To the extent that peaks, or pollution above certain critical levels, rather than averages are pertinent measures of risk to health, a number of observations in the course of each day may be needed. Sampling frequency is inadequate.[30] In the words of the third annual report of the Council on Environmental Quality: "All of the data . . . suffer from the problem of too few monitoring sites whose locations are determined by imprecise criteria. Unreliable instrumentation and inadequate frequency of measurement are also problems. It will take a long time to remedy these defects."[31]

Measurements of air pollution must be carried on over a long period of time as a basis for estimating all the important effects of air pollution on health. We do know that some effects occur with long lags—as much as twenty years for cancer. There are no reliable time series on relevant pollutants that go that far back.[32] Other effects may be related to prolonged or cumulative exposure, even though lags in some effects may be short. Of course, where cumulative exposure or long lags have to be considered, the problem of relating pollution to exposure is complicated by the mobility of the American population. Most studies simply do not deal with this issue, although a few confront it and attempt to cope with it by using length of residence in an area as a proxy for cumulative exposure.

Pollution Indices

People are exposed not only to carbon monoxide or sulfur dioxide, but to a combination of pollutants that varies between areas and from time to time. When we speak of air pollution, therefore, we imply some mix of pollutants. An index of the level of air pollution must consequently deal with three issues: the pollutants to be included, the relative weight to assign the various pollutants, and the choice of observations: should they be averaged, hourly, daily, weekly, or over some other period, or should they be the pollution peaks, again, measured over what time periods?

Several indices have been used to estimate trends in air quality, including the Mitre Air Quality Index, Mitre Extreme Value Index, and the Oak Ridge Air Quality Index.[33] Since all of them relate measured air quality to EPA secondary air quality standards and consider the same pollutants—CO, SO_2, suspended particulates, photochemical oxidants, and NO_2 (the Extreme Value Index excludes NO_2—they yield similar results. They give the same weights, implicit in EPA standards, to various pollutants. The differences are in the weight given to the one-hour vs. the annual mean vs. intermediate periods, since the EPA has different standards for these time periods.

> All share at least four shortcomings: first, they do not adjust for the number of people exposed to given levels of air pollution. That is, they are indices of pollutant levels, not of the extent of population exposure. Second, they do not distinguish between changes in air quality due to control measures and changes due to natural meteorological fluctuations such as the amount of rain or number of inversions during the year. Third, they do not take account of interaction among different pollutants. Fourth, all the indices are based on a direct linear ratio between the ambient levels and a given standard, even though the relationship between pollutant levels and the damage caused is, in many cases, nonlinear.[34]

Perhaps the most serious defect of the indices for day-to-day readings is that they incorporate hydrocarbons and NO_x, rather than the far more harmful and irritating compounds formed in the atmosphere by complex and little-understood reactions of these pollutants in the presence of sunlight. Readings on HCs and NO_x simply do not reflect the concurrent concentration of the reaction products. Sulfur dioxide is less harmful than the sulfates to whose formation it may contribute. As to particulates, the size of the particles, as well as their chemical composition, may be more important than their volume. Thus, increases in the irritant quality of the air may be accompanied by declines in the index, and vice versa, and the same reading can indicate substantially different threats to comfort and health. For instance, pollen may cause more discomfort and greater economic loss than most of the pollutants indexed.[35] Although of natural origin, the amount and nature of vegetation in and around urban areas is very much in the hands of man. Enough people are allergic to ragweed to justify the pollen counts made in some places during the critical season.

Any index of pollution giving relative weights to different pollutants, or to pollution levels at various times (peaks, averages), implicitly includes estimates of the relative harmfulness of different pollutants and of different levels of concentration of each pollutant. Several estimates of the relative severity of different pollutants are given in table 2-2. None of them includes oxidants, and their greatest disagreement is on hydrocarbons, which are relatively harmless in themselves, but are oxidant precursors. Since EPA primary standards for different pollutants are expressed in terms of maximum concentrations for entirely different time periods, ranging from one hour to an annual average, they cannot be compared directly with these severity estimates.

The reported indices are only a few among many. A survey of the thirty-five largest cities in 1975 found that they were using thirty-two different indices.[36] As one consequence of this confusing situation, the Federal Interagency

Table 2-2
Air Pollutant Severity Factors

	Severity Factors				
Pollutant	*Caretto-Sawyer*	*Walther*	*Babcock 1970*	*Babcock-Nagda 1972*	*RECAT Committee*
Carbon monoxide	1	1	1	1	1
Hydrocarbons	62	124	2	10	60
Nitrogen oxides	44	22	78	24	100
Sulfur oxides	27	15-21	28	29	120
Particulates	38	21-37	107	52	133

Source: Lyndon R. Babcock, Jr. and Niren L. Nagda, "Cost Effectiveness of Emission Control," *Journal of the Air Pollution Control Association*, vol. 23, no. 3 (March 1973), p. 174. Reprinted with permission.

Task Force on Air Quality Indicators was established in 1975 and developed its own suggested index.[37]

The severity estimates imply a linear relation between pollution levels and health effects that does not always obtain. They also imply that the effects of several pollutants combined is equal to the sum of the effects of each independently. In fact, the relations may not be additive, but multiplicative (synergistic), or even substitutive. For example, SO_2, a reducing agent, will lower the atmospheric concentration of oxidants such as PAN (peroxyacyl nitrates—an irritating component of smog) and ozone. Substitution implies an upper threshold to health damage. Russell Train, in his report of October 1972, stated that "in many cases the scientific knowledge necessary to properly weight the components of an index is lacking."[38] "It appears . . . that the nation is embarked on an air-pollution-control program of enormous scope, complexity, and cost with little measure of relative harmfulness to health of the several pollutants being considered."[39]

Indoor Versus Outdoor Air Pollution

The dominant stress of the Clean Air Act and of the EPA is on the outdoors. Concern with indoor air refers to industrial facilities whose workers may be subject to excessive exposure to dangerous substances connected with their work. Many people, however, work in offices rather than factories. Most people spend a large part of their time indoors, and even among those who work, more time is spent indoors at home than indoors at work. Typically, pollution is lower indoors than outdoors, and the differential is a function of the rapidity of air circulation between outside and inside.

There appears to be a linear relation between particulate concentrations indoors and outdoors, with indoor particulates smaller in size. It makes a large difference whether windows are open or not. Climate is a major influence on the relation between indoor and outdoor pollution.[40] A study of six air pollutants as a function of air circulation found that inside oxidant levels were two thirds of outside levels with rapid circulation, but dropped to near zero with little intake and slow circulation. Particulate levels depended on velocity or air movement. Indoor levels of oxides of nitrogen and carbon monoxide, however, were much more stable than oxidants, whereas PAN occupied an intermediate position.[41]

Not all indoor pollution is brought in from the outdoors. Household activities, particularly combustion in the kitchen and the furnace, generate pollutants in concentrations that may exceed outdoor pollution. Pressurized spray containers are a factor. High levels of mercury vapor concentration have been found in homes recently painted with latex paint, which contains mercury compounds to prevent the growth of fungus, and in some doctors' offices.[42] Perhaps the worst indoor pollutant is cigarette smoking, which produces large amounts of CO as well as NO_x and numerous hydrocarbons, including some that are highly carcinogenic.[43] The harmful by-products of smoking are more concentrated indoors than outdoors, increasing exposure to both the smoker and the nonsmoker. The EPA does not deal with this form of air pollution.

3 Effects of Air Pollution Exposure on Health

In order to understand the impact of man-made air pollution on health, we must know many complex reactions. First is the relation between pollutant emission and ambient air pollution. Second is the relation between ambient air pollution and exposure of the population. Third is the relation between exposure and impacts on health.

Possible Relations Between Pollution Exposure and Health

In designing better policies to protect health, it is not enough to know that there is or is not a relation between an air pollutant and health. The form of this relation must be known. For each pollutant capable of causing damage to health, the following questions need to be answered:

1. At what levels of concentration does damage to health begin to occur? Is there a minimum critical threshold of pollution below which a pollutant can be ignored as essentially harmless?

2. Above any threshold level, what is the relation between concentration of pollutant and damage to health? This question may have to be answered in two parts, to consider the increasing proportion of the population that is affected and the increasing harm to individuals affected. Specifically, is the relation linear?

3. What is the relation between duration of exposure to any given concentration of pollutant and damage to health? This in turn involves a number of considerations: the existence of a threshold below which there are no effects; linearity or nonlinearity of effects as a function of level of pollution and of the duration of exposure. At issue is the reaction of the human organism to prolonged or repeated exposure. Does the body build up a tolerance to pollutants so that chronically exposed populations suffer less damage than those who are first exposed?

4. What is the effect on health of the distribution of a given average pollution over time? Does the relation between pollution and health depend only on cumulative exposure, irrespective of variations in its concentration over time? Or do peaks have disproportionate impacts on health, so that the greater the amplitude of fluctuation in pollution levels, the greater the damage? Depending on the

answer to this question, policy would focus either on reducing cumulative exposure or on leveling peaks.

5. What are the lags between harmful pollution and observed damage to health? Here it is important to distinguish between different kinds of morbidity, because they have widely different lags, from near-instantaneous allergic reactions to decades-long lags in the case of some cancers. Are the costs of pollution immediate or deferred?

If the effects may be long delayed, then currently observed effects may be the consequence more of past than of present concentrations of pollutants. Thus, correlation analyses of health statistics and measures of air pollution over short time periods may overlook much of the cause and also neglect much of the effect. On the other hand, it is well known that long time series involve severe problems of isolating particular cause–effect relations from among a host of variables and interactions (including, for instance, migration).

6. How persistent are the effects, after cessation of their cause? Are they reversible or self-limiting, or are they enduring or progressive? Again, it is necessary to distinguish between different types of morbidity which may be attributable to the same pollutant. The answer would indicate the long-term cost of pollution.

7. In fact, individuals are typically exposed to a combination of pollutants. Is the effect of various combinations equal to the sum of the effects of separate pollutants, or is there a multiplicative effect, such that the combination is worse than its separate components? Or is there a subtractive effect, so that different pollutants interfere with and counteract each other, with less impact on health than the sum of the effects of the separate pollutants?

There are two ways of approaching the effects of air pollution on health: from the standpoint of widespread air pollutants that pose some threat, or from the standpoint of the various harmful effects resulting from air pollution. Since regulatory policy is expressed in terms of pollutants, this will be the approach followed below.

Indicators of Health Effects

Measuring health and illness is no less complex than measuring air pollution and exposure to it. At one extreme, mortality (or rather premature or excess mortality, since we all die anyway) is a measure of damage to health. Less severe effects may be indicated by hospitalization rates, hospital bed-days, or doctors' visits, which may be classified by type of ailment. Even milder effects may be reflected in absenteeism from work and school. These indicators are probably ordered in terms of decreasing reliability. It is hazardous to associate absenteeism, for instance, with health, much less with air pollution-induced health problems.[1] Even with regard to hospitalization rates and medical reports, there is a lack of uniformity in disease nomenclature as well as in data collection procedures.[2]

All of these indicators overlap. There is double counting: some patients die; some absentees visit doctors, are hospitalized, and may die. Other people, who are as severely afflicted as many counted by some indicators above, just soldier on, and the only way to obtain evidence of their symptoms is to elicit self-reports, a less reliable sort of evidence than hospital or coroners' statistics.

At every level of severity, a variety of ailments associated with air pollution exposure can be identified. Essentially no work other than economic measures of cost and days lost has been done on an index of morbidity. We have no index assigning relative weights to a visit to the doctor, a day in a hospital, and a premature death, nor to different types of disease. The effects on health associated with air pollution are as diverse as the causes themselves.

Major Air Pollutants and Health

Studies of the effects of air pollution on humans are of two kinds: those that observe physiological counterparts of exposure to air pollution (effects on lung capacity of carbon monoxide in the blood), and those that observe mortality or indicants of morbidity, such as hospitalization or absenteeism. Only the second refers to effects on health; the first kind notes what may or may not be intervening variables caused by air pollution, and which in turn may affect health. It should be noted that most studies focusing on morbidity and mortality relate air pollution and health, not exposure and health.

Carbon Monoxide

Two kinds of health effects are attributed to CO, very short term, and long term. Exposure to high concentrations of CO raises the level of carboxyhemoglobin in the blood. One argument for harmful short-run effects posits that high blood concentrations lead to diminution of attention and accuracy in problem-solving tasks, reduced sensory discrimination and visual acuity, and on to more traffic accidents and damage to life and limb. The evidence of such physiological effects of urban concentrations of CO in areas of congested traffic is by no means conclusive. Other studies have failed to find evidence of such effects, even at very high levels of CO concentration in the air and in the blood.[3] Studies confirming a positive correlation between CO concentration and high accident rates and mortality fail to allow for the fact that such concentrations occur in periods of heavy and congested traffic, when opportunities for accidents are high and access to medical care slow. Thus, the impact of CO on accident rates and mortality is not proven.[4]

Studies of small groups subjected to unusually high levels of CO, such as traffic policemen and tunnel workers, have found evidence of respiratory effects: asthma, bronchitis, and reduced respiratory function.[5] Other studies have failed to find such effects, even at concentrations many times higher than EPA standards.[6] Since exposure to automobile exhaust includes pollutants other than CO, even firm evidence of correlation between CO and respiratory conditions does not establish the causative role of CO. Nor is it clear that the effects are exclusively the result of short-run exposure to high concentrations; they may be the result of often-repeated exposure.

Short periods of exposure to very high concentrations of CO can precipitate or hasten the death of individuals with severely impaired heart or respiratory functions. The high level of CO in the blood imposes an additional strain.[7] Individuals suffering from angina pectoris or oxygen insufficiency from other causes find their symptoms aggravated by very high levels of blood carboxyhemoglobin—levels much higher, however, than those associated with congested urban traffic, or with the EPA ambient air quality standard.[8]

Most studies distinguish between smokers and nonsmokers. It is clear that smoking is more important than air pollution in raising blood carboxyhemoglobin levels, and that many smokers exhibit levels far above those that some studies associate with pathological effects.[9]

The short-run effects are fully reversible. Blood carboxyhemoglobin levels drop to normal in a matter of hours after termination of exposure to high concentrations of ambient CO. In the longer run, it appears that effects are mitigated by adaptation of the human organism to prolonged exposure to high levels of CO. This ability to adapt is one reason why different experiments yield such sharply conflicting results.[10]

A relation between prolonged CO exposure and disease of the heart or arteries is suspected, but not established.[11] Animal studies suggest the possibility of damage to the central nervous system from chronic prenatal exposure to CO.[12]

Sulfur Dioxide and Other Sulfur Compounds

Most episodes of fogs causing substantial excess mortality have involved very high and prolonged concentration of SO_2 as a result of temperature inversions, reaching ten to forty parts per million (ppm) in the Meuse Valley, Belgium, in 1930, and over one part per million in Donora, Pennsylvania, in 1948, and in the London fogs of 1952 and 1962.[13] (The EPA maximum twenty-four hour concentration is 0.14 ppm.) But in every case, high levels of particulates and humidity resulting in sulfuric acid mists were also present. Prolonged inversions allowed atmospheric chemical reactions to produce high levels of toxic secondary pollutants, sulfates in particular.[14]

While the role of SO_2 in isolation in contributing premature death of vulnerable individuals may not be established, even at high concentrations of one ppm, it is clear that in combination with particulates and high humidity it and/or the sulfates into which it is transformed are a threat to health and life.[15] At lower concentrations, often well above the 0.14 ppm standard, evidence on excess mortality is contradictory.

Buechley and others found mortality in New York City 2 percent greater than expected on days with SO_2 levels above 0.2 ppm, and 1.5 percent below expected levels when the SO_2 concentration fell below 0.01 ppm during the period 1962-66.[16] The coefficient of haze, however, was as good a predictor as SO_2. Similar results were obtained for Philadelphia. But Schimmel and Murawski have questioned the relation of sulfur dioxide and mortality in New York City.[17] They noted that average daily levels of sulfur dioxide dropped significantly over the past decade (1963-72), but could find no corresponding decrease in deaths associated with air pollution. Smoke shade, which did not change over the period, could account for nearly all excess deaths.

Heimann, studying an air pollution episode in Boston in 1966, found no relation between pollution levels and mortality. Sulfur dioxide levels of 0.11 per million were only one quarter to one fifth of those in New York during crises in 1962 and 1965, a period over which Hodgson observed a positive relation between monthly air pollution and mortality.[18] McDonald and Schwing, studying age-adjusted mortality across sixty standard metropolitan statistical areas (SMSAs)–urban areas with a central city of at least 50,000 people–found sulfur dioxide quite significant in explaining differential mortality.[19]

Lave and Seskin studied variations in mortality across 117 SMSAs in 1960 and replicated their study in 1961 and again in 1969 (with 112 SMSAs).[20] The biweekly minimum sulfate level proved to be the most significant pollution variable. Variations in death rates from total cancer, digestive system cancer, and cardiovascular disease were closely associated with sulfate concentration, but respiratory disease mortality was not. A linear relation proved at least as good as any other. When daily variations in death rates and pollution were examined for each of a number of cities in the mid-1960s, SO_2 concentration in Chicago was positively correlated with death rates, but there was no consistent effect of pollution on mortality in Denver, Philadelphia, St. Louis, or Washington, D.C., cities with much lower levels of pollution than Chicago. This finding is not consistent with a linear relation over the entire range of pollution observed.

There is little doubt that SO_2 concentrations such as are frequently found in some of our large industrial centers do irritate the respiratory tract and increase the incidence of respiratory disease. Durham, surveying colleges in the Los Angeles basin, found pharyngitis, bronchitis, tonsillitis, colds, and sore throats associated primarily with photochemical oxidants, sulfur dioxide, and nitrogen dioxide.[21] Air pollution differentials between Los Angeles schools may be responsible for a 16.6 percent increase in these respiratory symptoms.

A study comparing accidental deaths in St. Louis and Winnipeg found a much higher incidence and a much greater severity of emphysema in St. Louis.[22] However, as in the case of the New York and Boston studies, SO_2 was not the only pollutant attaining high levels of concentration in St. Louis.

There is evidence that high concentrations of SO_2 (always with other pollutants) are associated with reduced respiratory capacity, which in turn may be expected to be conducive to more severe respiratory disease. There is some disagreement on the duration of these effects. Stokinger and Coffin conclude that sulfur oxides and sulfuric acid do not "give rise to chronic irreversible changes in the lung from repeated or continuous daily exposure at any attainable air pollution concentration."[23] On the other hand, Shy's study of schoolchildren in New York concluded that early exposure to high levels of air pollution for five to ten years (his index was SO_2 and particulates) resulted in prolonged decrement in ventilatory function.[24] Schoolchildren in Cincinatti, however, presumably exposed to much lower levels of pollution than those in New York, recovered their ventilatory capacity quickly. There is also evidence that prolonged exposure to SO_2 develops resistance to its effects.[25]

SO_2 is not the most abundant sulfur compound emitted into the atmosphere. Hydrogen sulfide (H_2S), predominantly a natural product of the decay of vegetation, is more abundant. It is harmless even at concentrations many times ambient levels. But it reacts to form SO_2 and sulfates, which are harmful. Its life in the atmosphere is very brief, less than one day.[26] Sulfur dioxide itself may be quickly transformed into SO_3, sulfuric acid (H_2SO_4), and sulfates, particularly ammonium sulfate. Measures of atmospheric concentration of any one sulfur compound may be poor indicators of the concentration of other toxic compounds, and this fact may help explain contradictory findings.

Nitrogen oxides

Nitrogen oxides directly affect health and are also transformed in combination with other air pollutants into harmful secondary pollutants, in particular photochemical oxidants. The most common oxide, N_2O, produced predominantly by natural processes and harmless, contributes to the formation of harmful secondary pollutants. Man-made emissions, predominantly NO_2, on the other hand, are harmful in high enough concentration as well as being constituents of secondary pollutants. Since the original NO_x and the secondary pollutants coexist, although not in any fixed proportion, associations of NO_x and morbidity may mistakenly attribute pathological symptoms to NO_x which are properly the consequence of secondary pollutants. Nevertheless, it is true that reduction in NO_x is likely to reduce the secondary pollutants as well.

Nitric oxide (NO) has effects similar to those of CO: it causes oxygen deprivation, because of its superior affinity for hemoglobin. It has no other harmful effects and is not found in sufficient concentration to be of concern. NO_2, on the other hand, is an irritant to respiratory tissues and causes chest pains, pulmonary edema, and death. It also appears to reduce the body's resistance to infection.[27] But findings of the more severe consequences are based on exposures many times the peak ambient air concentration, or on animal studies. The concentrations to which smokers are exposed are suspected as a major factor in emphysema.[28]

Evidence that ambient air concentrations can have harmful effects on health is limited and equivocal. A study in Chattanooga found a correlation between NO_2 exposure and reduced respiratory function in children, and greater likelihood of acute respiratory illness.[29] The correlation applied to bronchitis but not to pneumonia or croup, or to hospitalizations.[30] The findings were not entirely consistent, and have been criticized.[31] Another writer has found no evidence of correlation between NO_2 concentration and hospital admissions or physicians' house calls in the United States.[32]

Lave and Seskin did a cross-sectional analysis of the sixty-nine SMSAs with data on NO_2 and nitrates (NO_3) in 1969. They found significant relations between NO_2 concentration and mortality. Of the five large SMSAs for which they studied daily variations in pollutant concentration and mortality, only Chicago revealed a significant relation between NO_2 and mortality, and only Philadelphia between NO and mortality. The findings for nitrogen compounds, however, were less consistently significant than those for sulfates, SO_2, and particulates, and contributed little to the explanatory power of the regressions.[33] A single report indicates a strong association between NO_2 and long-term mortality from hypertensive heart disease.[34]

There is no question that NO_2 can have these effects in sufficient concentration, but there are doubts that ambient air concentrations found in congested urban centers are sufficiently high to produce these symptoms. Los Angeles reports 80.4 percent of the person-illness attributed to NO_2 for the United States, suggesting a highly concentrated and nonlinear relation.[35] Laboratory studies, primarily with animals, that have employed much higher concentrations of NO_2 than are found in ambient air, indicate that the effects are reversible once exposure is terminated, and that individuals under prolonged or repeated exposure build up tolerance to the harmful effects of NO_2.[36]

Thus, the main cause for concern appears to be the contribution of nitrogen oxides to the formation of more dangerous chemicals in the air. NO_2 is converted in the atmosphere into nitrates and nitric acid that precipitate in acid rains, with adverse effects on plant and animal life. NO_x, in the presence of sunlight and sulfur oxides, is a key contributor to photochemical production of sulfates in the air. NO_x produced by high-flying aircraft depletes stratospheric ozone which shields the earth from ultraviolet radiation, thereby increasing the incidence of skin cancer.

Hydrocarbons

Hydrocarbons have direct and indirect effects on health. The most prevalent, methane, largely of natural origin, is not harmful, but its breakdown produces large amounts of CO. The majority of other hydrocarbons are also harmless, but some of the aromatics, such as benzo(a)pyrene, are highly carcinogenic.[37] Of greater importance is the role of hydrocarbons in producing highly reactive oxidants when in combination with oxides of nitrogen and in the presence of sunlight. The resulting photochemical reactions, responsible for California-type smog, produce oxidants that are irritating to the eyes and respiratory tract, in particular ozone (O_3) and peroxyacyl nitrates (PAN).

Because the threat of hydrocarbons lies primarily in their contribution to the formation of oxidants, for which there is an air quality standard, it has sometimes been argued that an air quality standard for hydrocarbons is superfluous. One possible adverse trend, however, is the introduction of no-lead gasoline. In order to minimize the loss of octane rating, refiners are likely to change the composition to include a higher proportion of aromatic and olefin hydrocarbons, which are themselves harmful and which react more rapidly in the photochemical process of oxidant formation.[38]

Oxidants

The preponderance of the oxidants in the air are not emitted by human activities, but are the result of chemical action in the atmosphere, for which the man-made emissions of hydrocarbons and nitrogen oxides provide some of the ingredients. Typically, their life in the atmosphere is extremely short because they are highly reactive—the very characteristic that makes them irritating and possibly harmful to health. Thus, the concentration of oxidants does not depend closely on the concentration of hydrocarbons and nitrogen oxides, but on the speed with which these compounds are transformed photochemically into oxidants. This depends largely on sunlight, and on the rate at which they are dispersed, a function of wind velocity and temperature gradients. Constant sunlight, lack of wind, and temperature inversion provide the set of conditions under which oxidant concentration is maximized.

Peroxyacyl nitrates, among the major constituents of photochemical smog, are irritating to the eyes but otherwise not proven hazards to health. Ozone, on the other hand, is highly toxic, even in small concentrations. It causes dypsnea (difficulty in breathing), bronchitis, emphysema, and pulmonary edema. It reduces respiratory function. It weakens the body's defenses against bacteria.[39] But there is some question whether any of these effects accompany even the highest ambient air concentrations of ozone. Some evidence exists of correlation between ozone concentration and hospital admissions.[40] Ozone affects athletic performance and is correlated with auto accidents.[41] It increases the probability of asthma attacks among asthmatics.[42] A single report finds a significant association between O_3 (and hydrocarbon) concentration in Chicago, Philadelphia, and Washington, D.C., and short-term excess mortality.[43] Its mutagenic potential is suspected. But the authors of *Air Quality Criteria for Photochemical Oxidants* conclude that "no convincing relationship was observed between short term variations in photochemical oxidants and (1) daily mortality or (2) hospital admissions."[44] Evidence from animal studies suggests that prolonged exposure can build up a long-lasting tolerance to ozone.[45]

Particulates

Particulates contain diverse constituents: soot, dust, pollens, lead, and some hydrocarbons. Therefore, air quality standards expressed in terms of particulate concentration are inappropriately characterized in terms of their effects on health. It is important to distinguish particulates by chemical composition and also by size.[46] It is unwise to generalize from findings for any one area to other areas whose particulate composition may be quite different, even though available measures of suspended particulates may not differentiate them.

Winkelstein, in a study of areas in and around Buffalo, found a strong positive association between suspended particulate levels and gastric and prostatic cancer and respiratory diseases.[47] He found a weaker association with cardiovascular diseases, and a strong suspicion of association between pollution and cirrhosis of the liver.

Lave and Seskin, in their cross-sectional analyses of 117 SMSAs in 1960 and 1961 and of 112 in 1969, found a significant correlation between particulate concentration and total mortality, and found particulates to be the pollutant most closely associated with the death of infants under one year of age. Cross-sectional time series analysis of twenty-six SMSAs corroborated these findings.[48]

It should be noted that particulates include polynuclear aromatic hydrocarbons, among them benzo(a)pyrene, already indicted as a cause of cancer, They also include some of the irritating secondary pollutants, such as sulfates and nitrates. It has been observed that the combination of particulates and SO_2 appears more damaging than either alone.[49]

Caprio and others have shown that children living close to traffic arteries in Newark have elevated blood lead levels, and they have singled out automobile

Table 3–1
Air Pollutants and Health: A Summary

Pollutant	*Symptoms*	*Consequences*	*Comments*
Carbon monoxide	Reduced visual acuity and sensory discrimination	Traffic accidents	Evidence not conclusive; effects short-term, reversible
	Reduced respiratory function	Asthma, bronchitis, aggravation of angina, arteriosclerotic disease	Increased tolerance with prolonged exposure. Evidence clear only for concentrations far in excess of EPA standards. Suspected long-term effects
Sulfur dioxide, sulfates	Irritation of respiratory system	Excess mortality	In combination with particulates and humidity, in concentrations above EPA standards
	Reduced respiratory function	Emphysema, bronchitis, other respiratory diseases	Some evidence of development of resistance with prolonged exposure
Nitrogen dioxide, nitrates	Irritation of respiratory tract	Bronchitis, other respiratory illnesses	Effects on health questionable at EPA standards
	Reduced respiratory function; reduced resistance to infection	Excess mortality; hypertensive heart disease suspected	Effects may be reversible; development of tolerance under prolonged exposure suspected, in single report
Hydrocarbons	A few are irritating to eyes, respiratory tract	Cancer	Most hydrocarbons are not themselves harmful, except aromatics and olefins; hydrocarbons are oxidant precursors (for example, benzo(a)pyrene)
Oxidants	Irritation of eyes, respiratory tract	Bronchitis, asthma, emphysema, difficulty in breathing	Effects on mortality questionable; lasting tolerance built up by prolonged exposure
Particulates	Irritation of eyes, respiratory tract	Excess infant mortality; respiratory disease; some types of cancer	Particulates include some sulfates, aromatic hydrocarbons, nitrates. Some, such as asbestos and nickel, cause cancer

exhausts as a factor.[50] However, lead content of human tissues has declined historically.[51] It is estimated that only one third of the lead assimilated by the body is inhaled; the remainder is dietary in origin.[52] It has not been demonstrated that lead inhalation in urban areas with congested traffic is a factor in pathological symptoms, such as anemia, of excess body lead.

Asbestos, usually regarded as a threat only to miners and workers in industrial activities involving it, causes asbestosis, a possibly serious lung condition. High ambient air concentrations in some urban areas, however, and the presence of asbestos fibers in the lungs of residents in some industrial centers suggests that the threat may be more extensive.[53] Its principal threat to health in urban air is as a causal factor in lung cancer.[54]

The effects of air pollutants on health are summarized in table 3-1.

Types of Illness Associated With Air Pollution

A summary of observed effects of air pollution by type allows us to focus on specific kinds of health damage and to identify airborne threats, including some substances for which there is no ambient standard or monitoring. (Some of the evidence has already been presented under the sections on different pollutants.)

Cancer

Since cancer becomes evident only years after exposure to carcinogenic pollutants, it is difficult to establish causation, and it is not always possible to isolate specific causes.

Lung cancer is strongly associated with ambient air pollution. Rates are higher in urban than in rural areas in the United States,[55] and increase with length of urban residence.[56] Within urban areas, death rates are positively related to exposure,[57] and resulting deaths occur at younger ages.[58] The urban-rural differential has been observed in a number of other countries.[59] Studies of British migrants to Australia, South Africa,[60] and New Zealand[61] reveal that migrants have a higher mortality rate from lung cancer than natives of these countries, but lower rates than nonmigrating British residents. The mortality rate of migrants to New Zealand varied directly with age at migration.

Lave and Seskin, reworking data from the United Kingdom on lung cancer mortality, obtain contradictory results.[62] When they rework British data on nonrespiratory cancer mortality, they find that stomach cancer is significantly related to indices of deposit or smoke, but intestinal cancer seems only marginally related. Cancer of the stomach, esophagus, buccal cavity, pharynx, larynx, bladder, and prostate, however, has been found by some studies to be associated with air pollution in the United States.[63]

Most studies relating air pollution and cancer did not isolate particular suspect pollutants. This is true of urban-rural gradients in mortality rates in particular. Even in cases where specific pollutants, including SO_2, coal consumption, particulates, and smoke, were used as indicants, other air pollutants were also present and not taken into account.

We do know, from laboratory studies and from studies of instances of industrial exposure to much higher concentrations than are encountered by the general population, that certain air pollutants are highly carcinogenic. This is true of benzo(a)pyrene, a product of combustion of hydrocarbon fuels, cigarettes, and petroleum refining. Carnow and Meier, using benzo(a)pyrene as an index of air pollution in a study across nineteen countries and forty-eight states, concluded that reduction of 60 percent in air pollution (by this index) might be expected to reduce pulmonary cancer deaths by 20 percent.[64] Menck, et al. found an apparent 40 percent excess male lung-cancer death rate in areas of Los Angeles close to refineries and chemical plants emitting benzo(a)pyrene.[65] Other air pollutants definitely indicated as carcinogens include asbestos, chromium, and nickel.[66]

Cigarette smoking is a more important cause of lung cancer than ambient air pollution. Studies of the relation between the amount smoked and cancer are essential to establish the contribution of air pollution. These studies indicate a linear relation between the incidence of cancer and the number of cigarettes smoked.[67] They also indicate the possible existence of an upper threshold: for very heavy smokers, urbanization and its associated air pollution does not appear to raise the incidence of cancer, whereas for other smokers urbanization does increase the probability of lung cancer.[68]

The risks of exposure to cancer-inducing pollutants are summarized in the statement: "It should be remembered that weak carcinogenic exposure has irreversible and additive effects and cannot be dismissed lightly as standing 'below a threshold of action'".[69]

Respiratory Disease.

The effects of pollution have been examined for respiratory disease in general, and for specific conditions, e.g., bronchitis and pneumonia. In the Buffalo area, a systematic relation was found between particulate concentration and death rates from respiratory disease and sulfur oxides.[70] A comparable study in Nashville, however, failed to find a relation between mortality and particulates, and found respiratory disease related to sulfur trioxide and haze and smoke, but not to SO_2 or dust fall.[71] Lepper and others found a positive relation with SO_2 in Chicago.[72] The findings of Shy et al.[73] and Pearlman[74] relating NO_2 in Chattanooga with respiratory and bronchitis morbidity, respectively, have already been reported.

When long-term effects were examined by correlating respiratory disease symptoms among Veterans Administration (VA) hospital patients with the size of city from which they came, no association was discovered.[75] Motley also failed to find physiological effects of breathing smoggy air on normal subjects, although the symptoms of emphysema victims were aggravated.[76] But Speizer and Ferris found that road tunnel employees in Boston with more than ten years' exposure to automobile exhaust in the tunnel had a higher rate of chronic respiratory disease than employees with shorter exposure.[77]

Numerous studies in the United Kingdom have found associations between pollution (usually measured by sulfur dioxide and smoke or particulates) and respiratory illness.[78] Urban-rural differences in respiratory disease, and differences between British and American occupational groups, were attributed to different exposure to air pollution. Differences were also found in respiratory function.[79] A study in Italy[80] and a number of studies in Japan[81] also associated respiratory morbidity and mortality with air pollution.

In Norway it was found that differences in respiratory disease and function related to exposure to sulfur dioxide disappeared after age of fifty.[82] A number of the studies finding striking correlations between pollution and disease were of young children. At least one found higher associations for very young children, especially those under one year of age, and lower ones for older children.[83]

Significant correlations have been found between bronchitis mortality in the United Kingdom and various indices of air pollution (sulfur dioxide, smoke, sulfate deposits).[84] Mortality has also been related to urbanization[85] and to duration of exposure.[86] Lave and Seskin, analysing the data reported by Stocks and Ashley, found a positive and highly significant correlation between bronchitis mortality and particulate deposit indices, smoke, and sulfur dioxide concentration, and a moderately significant correlation between pneumonia mortality and smoke indices.[87] The effects of variation in pollution on differences in mortality were large, in extreme cases by a factor of more than three. Significant correlations with various indices of air pollution have been found for particular occupational groups, mainly in the United Kingdom.[88] The correlates have included bronchitis and other respiratory diseases, absence from work due to bronchitis, and reduced respiratory function.

Yoshida correlated chronic bronchitis with SO_2 and dust fall in Japan.[89] Cederlof, in his study of same-sex twins in Sweden, found bronchitis symptoms more prevalent among urban than rural residents.[90] But Stocks, in a study of bronchitis mortality across twenty nations, found it significantly correlated with solid fuel consumption only for the United Kingdom.[91]

As to pneumonia, Mills found very large variations in death rates in Pittsburgh and Cincinnati correlated with soot fall.[92] There was an inverse correlation with elevation, presumably a measure of local variations in pollution and exposure. Death rates in Chicago were also correlated with soot fall and SO_2.[93] In the United Kingdom, correlations were found between pneumonia mortality and smoke, fog, sulfate deposits, SO_2, and fuel consumption[94]; and between tuberculosis mortality and fuel consumption and SO_2.[95]

Ishikawa and others, as already noted, found much higher incidence and greater severity of emphysema among accident victims in St. Louis than in Winnipeg.[96] Sultz and others found an association between particulates and incidence of hospitalization for asthma and eczema among children in Erie County, New York.[97]

Lave and Seskin examined total respiratory mortality and five diseases of the respiratory system. Only tuberculosis exhibited a statistically significant relation to air pollution. However, they warn that the small number of deaths from specific respiratory diseases to be expected in any one SMSA makes it very difficult to perform a satisfactory statistical test.[98]

Cardiovascular Disease

Manos and Fisher found high positive correlations between air pollution and arteriosclerotic heart disease and myocardial degeneration.[99] A consistent association was found between cardiovascular morbidity and pollution (SO_2, soiling) in Nashville.[100] As to mortality, association with particulates was found for females, but not for males.[101] Hodgson found a positive association between mortality and pollution in New York over a period of two and a half years.[102] Several studies have established the higher cardiovascular death rates in urban than in rural areas, and in metropolitan than in nonmetropolitan urban areas, without establishing a relation with air pollution.[103]

Gardner's study in the United Kingdom proved inconclusive, with most of the results not statistically significant.[104] On the other hand, a Swedish study of same-sex twins found a higher prevalence of angina pectoris in urban than in rural areas.[105]

Table 3-2 summarizes the association between particular diseases and various air pollutants.

The Trade-off Between Pollution and Health

To quote the conclusion drawn in the *National Emission Standard Study:*

> Cause-effect relationships for the pollutant levels and health effects . . . have by no means been scientifically established. The health effects described for each level are those that would be expected if the same conditions present in the hypothetical model were found by the investigators who performed the studies from which these tables were developed. While it is unlikely that the same conditions would be found, it is clear from carefully documented studies that the potential for health hazards does exist in the various exposure situations described.[106]

Table 3–2
Diseases Associated With Air Pollution

Disease	*Associated Pollutant*	*Findings*
Total cancer; cancer of the respiratory system; Cancer of the digestive system	Sulfur dioxide, sulfates, particulates, benzo(a)pyrene, asbestos, ambient air pollution; also numerous other agents whose incidence is localized, such as nickel, beryllium, chromium	Numerous studies find an urban-rural differential without specifying particular pollutants
Respiratory disease mortality and morbidity	Particulates, sulfur dioxide, nitrogen dioxide, smoke, auto exhaust	Some studies find no effect on normal subjects. Greatest effect is on young children. Numerous studies find an urban-rural differential, and a differential associated with length of urban residence
Bronchitis	Same as above	Related to duration of exposure
Pneumonia	Soot fall, smoke, sulfur dioxide, sulfates	––
Tuberculosis	Sulfur dioxide, particulates	Few studies
Emphysema	Ambient air pollution, nitrogen oxides	Contrast of St. Louis and Winnipeg. Numerous studies find aggravation of symptoms among emphysematics
Cardiovascular morbidity and mortality	Ambient air pollution, sulfur dioxide, sulfates, particulates	Some studies prove inconclusive, or find no significant relation. Aggravation of symptoms among sufferers of angina pectoris. Differential rates found between urban and rural residents and between metropolitan and urban residents.

Lave and Seskin arrive at rough quantitative estimates of the relation between air pollution and excess mortality. In their 1960 analysis of 117 SMSAs, a 50 percent reduction in sulfates and suspended particulates is associated with a 4.7 percent reduction in total mortality; for 1969, the latter figure was 5.8 percent. In their analysis of daily variations in pollution and mortality for Chicago, a 50 percent decline in SO_2 is associated with a 5.4 percent drop in mortality.[107]

Although not too much stress should be placed on the precision of these findings, given the data limitations previously discussed, Lave and Seskin did attempt to avoid the shortcomings of most other studies. They tried to control for the effects of other variables, both socioeconomic and environmental, that might affect the results. And, unlike many other studies, there is no problem of sample size for broad classes of disease. It should be noted, however, that they only studied mortality, assuming a relation to morbidity.

They also found that a linear relation between pollution and mortality appeared to be as good as any other relation.[108] This conclusion cannot be regarded as settled, and is not entirely consistent with some of their own results. The range of their observations was limited. Evidence that the effects of diverse pollutants are not additive includes the chemical interaction between pollutants and a number of laboratory studies on animals; the evidence is circumstantial, but sufficient to keep the question open.[109] The linearity of effects is in contradiction to efforts to establish thresholds, and to numerous claims, however at variance with one another, that there are thresholds below which no damage to health (and therefore, a fortiori, no excess mortality) can be expected. There is some evidence to support the linear relationship too. Cohen found no difference in lung function between residents of cities with the same average pollution, but quite different peaks.[110]

The most important issue, in terms of the costs of air pollution and the benefits of reducing it, is the foreshortening of life expectancy among those whose death is associated with air pollution. Whether this is a matter of days, weeks, or years makes all the difference in the world. The evidence for compensatory drops in mortality following termination of brief episodes of acute air pollution[111] and the evidence for the absence or partial nature of such compensatory effects[112] (indicating that mortality is not simply acceleration of death among the dying) are not necessarily contradictory. The latter evidence is not limited to the aftermath of acute episodes, but includes fluctuations of much lesser magnitude. Both effects presumably occur, and the factual question which cannot be answered conclusively at this point concerns their relative importance.

There have been hundreds of studies on the relation between air pollution and health. Most of them find a positive association between pollution levels and incidence of various illnesses or pathological symptoms. The reliability of the association is subject to the previously discussed qualifications on the very measurements which are associated. There is the further problem of attribution, of assigning causality to air pollution. There are many other possible causes of observed differences in rates of disease or mortality, including individual differences between the members of groups experiencing different rates (age; sex; race; ethnic group; behavioral habits, including smoking and diet) and environmental differences between different areas or time periods (climate, water supply, domestic heating practices and facilities). "We have already shown how difficult it is to prove that air pollution causes ill health, and it is even harder to quantify the nature of the relation."[113]

Although no one study holds constant all the variables that might explain observed associations between pollution and health, each is held constant by some studies. Thus Lave and Seskin, in the most complete survey of findings to date, conclude that the studies in their entirety do establish causality beyond any doubt, whatever criticisms may be directed at any particular study. The evidence is strongest for sulfur dioxide and particulates, and especially for the combination of the two. Smoking, however, may do more harm than all other sources of air pollution combined. The evidence against nitrogen dioxide and carbon monoxide is much weaker, and most hydrocarbons can only be indicted as accessories to the photochemical production of oxidants. We remain unsure about the minimum threshold of damage to health. We are quite ignorant about the relation between levels of air pollution and frequency or severity of symptoms.

If scientists are not yet agreed on the threshold at which health damage can be expected for any pollutant, they certainly cannot agree on the dose-response function. The difficulties are complicated by the facts that exposed individuals appear to adapt to pollutants, effects often take time, some effects are partially or totally reversible, and individual pollutants interact. Furthermore, different pollutants have qualitatively different kinds of effects: some are eye irritants, others cause cancer. Thus, it is not possible to calculate trade-offs in estimating dose-response relations when more than one pollutant is involved. All that can be said is that most pollutants appear to have no harmful effects at low concentrations, but that carcinogens may have relatively low minimum safe thresholds. Most pollutants at most concentrations have effects that are reversible.

The question of thresholds, linearity., and reversibility of effects refer not to particular pollutants as such, but to particular types of ill effects. Possibly there is no minimum threshold for cancer, whatever the carcinogenic agent; any amount increases the probability of cancer. Likewise, cancer is not reversible by discontinuance of exposure to pollutants. On the other hand, emphysema implies a certain minimum physiological damage; it does not progress, like cancer, without further exposure to pollutants. Hence a threshold appears to exist. The same is true of black lung, silicosis, and other dust-induced lung ailments, which require a large minimum cumulative exposure before they can be diagnosed, and which persist once established. At the other extreme, reduced oxygen transport in the blood, whether caused by CO (as it usually is) or by other air pollutants, such as NO, is quickly reversible once exposure is terminated.

What constitutes an unacceptable concentration depends on the effect of high concentrations on the health of the exposed population. But it also depends on variations in sensitivity among the population. A concentration which will have no perceptible effect on the health of most people may have severe effects on some, perhaps very few, of those exposed. The dose-response relation needs to distinguish between the reactions of given individuals to varying exposure and the distribution of sensitivity among individuals within a population.

4 The Regulation of Air Pollution

Chapter 3 dealt with available evidence on the relation between air pollution and health. This chapter is concerned with federal legislation and its implementation –particularly the Clean Air Act Amendments of 1970. Are national air quality standards and pollutant emission restrictions appropriate, in the light of our knowledge of the health implications of air pollution? Are the methods for achieving the desired air quality optimal? What criteria can be recommended for setting standards and defining methods of attaining them in our present state of ignorance and uncertainty?

The Air Quality Act of 1967 required the Department of Health, Education and Welfare (HEW) to develop air quality criteria, and the states to develop plans to achieve them. Before these tasks had been completed, the Clean Air Act Amendments of 1970 were legislated, setting deadlines for the establishment of ambient air quality standards and for the preparation of implementation plans by the states. Meanwhile, the National Environmental Policy Act of 1969 required that environmental values be weighed in all federal governmental decisions.

The 1970 amendments directed the EPA to establish national ambient air quality standards, national emission standards for new stationary sources of air pollution and dangerous pollution agents, and national motor vehicle emission standards. Authority was provided to regulate the composition of motor vehicle fuel and emissions from aircraft.

Air Quality Standards

Air quality criteria were developed for the six air pollutants that are the main concern of this report: CO, SO_x, NO_x, hydrocarbons, oxidants (usually measured as ozone), and particulates. In conformance with the Clean Air Act Amendments, the EPA promulgated national primary and secondary ambient air quality standards for each air pollutant. (See table 4-1.) Primary standards are those that must be attained and maintained in order to protect the public health, allowing for an "adequate" margin of safety (see section 109b of the Clean Air Act). Secondary standards are those considered necessary to protect the "public welfare" from any known or anticipated adverse effects associated with the presence of air pollutants.

The law does not define the term "public health," but the Senate report on the bill mentions "particularly sensitive citizens such as bronchial asthmatics and emphysematics who in the normal course of daily activity are exposed to the ambient environment."[1] It states: "Ambient air quality is sufficient to protect the health of such persons whenever there is an absence of adverse effect on the health of a statistically related sample of persons in sensitive groups from exposure to the ambient air." (It does not, however, intend to include individuals who are so sensitive as to be dependent on a controlled environment—for example, newborn infants.)

The term "welfare" is clarified in Section 302(h) of the act, which states:

> All language referring to effects on welfare inc'udes, but is not limited to, effects on soils, water, crops, vegetation, man-made materials, animals, wildlife, weather, visibility, and climate, damage to and deterioration of property, and hazards to transportation, as well as effects on economic values and on personal comfort and well-being.

By 1975, when the national air quality standards established in 1970 were to be attained, most of the 247 air quality control regions still fell short of one or more of the standards, and some regions are still years away from meeting all the standards.[2] Had no new facilities emitting significant amounts of pollutants nor major additions to existing polluting activities been allowed, the regions in violation of the standards would have been compelled to stop growing, and in some cases would have been doomed to decline. Yet these regions accounted for a large proportion of industry and of the urban population of the country. The EPA responded to this dilemma with a compromise. So-called "offset" rules were adopted, permitting new pollutant sources provided that offsetting reductions in pollution were made. In some cases the EPA required offsetting reductions of twice the amount of pollution introduced by the new source. Furthermore, a region could be exampted from this trade-off requirement if they were making reasonable progress toward meeting standards. In 1977 the Congress extended the deadline for meeting standards for most pollutants to the end of 1982, and to 1987 for photochemical oxidants. It specified that no new industrial plants can be built after July 1, 1979, in any area not having attained air quality standards, unless the state has adopted an EPA-approved control plan assuring full compliance by the new deadline.

In March 1976 the Natural Resources Defense Council won a suit to compel the EPA to establish ambient air quality standards for lead. Since the catalytic converters installed in new cars since 1975 to reduce NO_x emissions require lead-free gasoline, levels of airborne lead should be declining. The standard proposed by the EPA for 1982 is 1.5 mg of lead per cubic meter of air. Ninety percent of lead in the air is believed to derive from gasoline consumption. The

Table 4-1
National Ambient Air Quality Standards
(micrograms per cubic meter)

Pollutant	*Primary Standard*	*Secondary Standard*
Sulfur oxides	80 (0.03 ppm)–annual arithmetic mean	60 (0.02 ppm)–annual arithmetic mean
	365 (0.14 ppm)–maximum 24-hour concentration[a]	260 (0.1 ppm)–maximum 24-hour concentration[b]
Particulates	75–annual geometric mean	60–annual geometric mean[c]
	260–maximum 24-hour concentration[a]	150–maximum 24-hour concentration[a]
Carbon monoxide	10,000 (10 mg) (9 ppm)–maximum 8-hour concentration[a]	Same as primary
	40,000 (40mg) (35 ppm)–maximum 1-hour concentration[a]	Same as primary
Photochemical oxidants	160 (0.08 ppm)–maximum 1-hour concentration[a]	Same as primary
Hydrocarbons	160 (0.24 ppm)–maximum 3-hour concentration (6 to 9 A.M.)[a]	Same as primary
Nitrogen dioxide	100 (0.05 ppm)–annual arithmetic mean	Same as primary

Source: As extracted from the *Federal Register,* vol. 36, no. 84, Friday, April 30, 1971, p. 8187, reproduced in U.S. Office of Science and Technology, *Cumulative Regulatory Effects on the Cost of Automotive Transport* (RECAT), Final Report of the Ad Hoc Committee (February 28, 1972), p. I–G2.

[a]Not to be exceeded more than once a year

[b]As a guide to be used in assessing implementation plans to achieve the annual standard

[c]As a guide to be used in assessing implementation plans to achieve the 24-hour standard

EPA will limit the lead content of gasoline to 0.5 grams per gallon after October 1979.

The EPA is also concerned with two other categories of pollutants. The first are pollution agents "whose presence, chronically or intermittently, in trace concentrations in the ambient air, either alone or in combination with other agents, cause or will cause, or contribute to, an increase in mortality or an increase in serious irreversible or incapacitating reversible damage to health." These are asbestos, cadmium, mercury, and beryllium. Second are those agents which are not emitted in such quantities or are not of such character as to be widely present on a continuous basis or readily detectable in the ambient air by present technology. These include arsenic, chlorine gas, hydrogen chloride, copper, manganese, nickel, vanadium, zinc, barium, boron, chromium, selenium, pesticides, and radioactive substances.

Motor Vehicle Emission Standards

The Clean Air Act Amendments of 1970 required that the emissions of carbon monoxide and hydrocarbons from light-duty vehicles manufactured during or after model year 1975 to be 90 percent below ceilings for 1970 vehicles. It also required that emissions of oxides of nitrogen for 1976 models and thereafter be 90 percent below the average of 1971 model light-duty vehicles. These standards and deadlines were to be met regardless of cost. On the other hand, the amendments required that stationary pollutant sources be controlled by application of the best available technology, taking into account the cost of achieving such control. The law provides for flexibility in the imposition of standards to abate emission of pollutants from new and existing aircraft, vessels and boats, railroad diesel engines, and trucks, buses, and other commercial vehicles.

Standards and test procedures were announced by the EPA in 1970 and 1971.[3] The 1975 model year standards were 0.41 grams per vehicle mile for hydrocarbons, 3.4 grams per vehicle mile for carbon monoxide, and 3.0 grams per vehicle mile for oxides of nitrogen. The 1976 model year vehicles were to meet the exhaust emission standard for oxides of nitrogen of 0.4 grams per vehicle mile. In 1973 the EPA postponed for one year the 1975 ceiling on emissions; an interim standard of 2 grams per mile was established. The Energy Supply and Environmental Coordination Act of 1974 extended 1975 interim standards for hydrocarbons and CO for one year, and gave the EPA administrator authority, exercised in 1975, to prescribe interim standards for 1977. In August 1977, Senate and House conferees agreed on a bill which again postponed, by two or more years standards initially intended to apply to 1975 model year cars. The 1977 model year standards were established at 1.5 grams of hydrocarbons per mile, 15 grams of carbon monoxide, and 2 grams of NO_x. For 1980, the permissible emissions drop to .41 grams of HC (the initial standard), 7 of CO, and 2 of NO_x. For 1981, CO is to attain the initial standard of 3.4 grams, and NO_x is to be limited to 1 gram per mile.

Also in 1977, the EPA established emission standards for motorcycles, to be reached in stages: starting with motorcycles built in 1978, allowable emissions were to be reduced each year until 1980, when they would represent a reduction of 54 percent in HC emissions and 49 percent in CO emissions. It adopted tougher emission standards for buses and heavy-duty trucks built in

1979 and later, reducing allowable hydrocarbon emissions by 50 percent, carbon monoxide by 12 percent, and nitrogen oxides by 21 percent.

Since 1974 the EPA has required jet airliners to be equipped with smoke reduction devices, and has forbidden pilots from dumping excess fuel after takeoff. Emission standards first set in 1973 for CO, NO_x, and hydrocarbons have been relaxed and delayed; they are to take effect in 1981.

The Energy Policy and Conservation Act of 1975, although intended to economize on gasoline rather than to reduce emissions, will have substantial impact on the latter. It provided that the average fuel requirement of new cars, expressed in miles per gallon, would be eighteen in 1978, nineteen in 1979, twenty in 1980, and twenty-seven and one half by 1985. In 1977 a standard of twenty-two miles per gallon was established for 1981, twenty-four for 1982, and twenty-six for 1983. President Carter proposed tax penalties on "gas guzzlers" and tax rebates on more economical cars. Congress rescinded the penalty, and postponed the rebate until the 1979 model year.

Emission Standards for Stationary Sources

Standards of performance for new stationary sources were announced by the EPA in 1971. The standards were for steam generators, portland cement plants, incinerators, nitric acid plants, and sulfuric acid plants. They covered emission limits for one or more of the four pollutants: particulate matter, sulfur dioxide, nitrogen oxides, and sulfuric acid mist for each source. The regulations included requirements for performance testing, stack gas monitoring, record keeping, and reporting.[4] In 1972 the EPA proposed revision of its regulations for new plants to allow emissions to exceed the standards set under certain operating conditions.

The EPA established emission standards in 1973 for seven additional stationary sources: asphalt concrete plants, petroleum refineries, petroleum storage tanks, secondary lead smelters, brass and bronze ingot production plants, iron and steel plants using basic oxygen furnaces, and sewage treatment plant incinerators.[5] It also relaxed SO_2 emission controls for utilities and nonferrous metal smelters.[6] It would allow polluters to rely on tall stacks, atmospheric conditions, and other factors to disperse emissions instead of installing expensive sulfur scrubbing equipment, but only under specified conditions that indicated that the need for such equipment would be infrequent. In 1976 it issued regulations sharply curbing the emissions from new copper, lead, and zinc smelters.

In recent years some major concessions have been made on emissions for coal-burning plants in order to encourage greater use of coal. In 1974 legislation provided that the Federal Energy Administration (FEA) must prohibit power plants (and may prohibit other major fuel-burning installations) from using oil or gas if conversion to coal is practicable, coal is available, and reliability is not impaired. The FEA designated eighty power plants in twenty-nine states as candidates for conversion to coal, of which the EPA has approved thirty-two so far. They will be given until 1979 to achieve standards. Those that have missed earlier deadlines are eligible for a three-year postponement. Also, in the interest of energy independence, the EPA has reviewed state plans for all 247 air quality control regions, and selected some thirty candidates for relaxation of particulate control requirements.

The EPA has been concerned primarily with stationary sources emitting a hundred or more tons of pollutant per year. It has not extended its mandate to include the many millions of heating and air-conditioning units burning hydrocarbon fuels which are installed in homes, apartment buildings, and commercial and industrial sites, and which collectively account for a significant share of pollutant emission. These are mainly space heaters and coolers, although stoves and hot-water heaters also contribute. The pollution control measures adopted wth regard to these important sources of sulfur oxides have been quality standards for coal and petroleum products, imposed by states and localities which set upper limits to the sulfur content of fuel.

Development and Land Use Control

The EPA has two different approaches toward clean air: it sets ambient air quality standards, and it prescribes emissions standards which are intended to enable the attainment of air quality standards, even in the most polluted areas. It has become clear, however, that the linkage between the two standards is not close, and that additional policies are needed to mesh them together. For instance, multiplication of emission sources in an area will reduce air quality, even though all sources meet emission standards. Thus, control over the number of new emission sources must be established to prevent air quality degradation. Such controls could eventually reorder the economic geography of the nation.

In June 1973 the Supreme Court, in a 4-4 decision, barred the EPA from allowing deterioration of existing air quality even in areas where air quality was far better than required to conform with national standards.[7] This "nondegradation" policy, whose specific implementation is uncertain (as the tie vote would suggest), may have far-reaching implications. In November 1974 the EPA adopted a threefold classification of areas: Class I, where air quality was to be protected at existing levels; Class II, where moderate changes would be permitted; and Class III, where major industrialization and growth would be allowed to the point where the area reached national air quality standards.

Even with the number of emission sources unchanged, and total emissions in a metropolitan area unchanged, mobile emission sources can change the patterns of concentration and dispersion in such a way as to create or aggravate local air pollution problems. Accordingly, in 1973 the EPA required states to consider the environmental impact of any new or modified facility that may reasonably be expected to interfere with the attainment or maintenance of a national air quality standard, such as major highways and airports, large regional shopping centers, major municipal sports complexes or stadiums, major parking facilities, and large amusement and recreational facilities.[8] If such "complex sources" of air pollution threaten to violate national air quality standards, states may be required to modify construction plans, or, if necessary, to stop them. However, implementation by the EPA was postponed repeatedly and, in 1975, indefinitely.

Such policies can eventually reshape the microstructure of most metropolitan areas. At issue here is the trade-off between air pollution averages and variations. Policies dispersing shopping or entertainment, for instance, may indeed reduce local pollution peaks, but at the cost of more car-miles in the metropolitan area, and therefore of higher average pollution levels.

National Trends in Emission and Air Pollution

Between the 1974 and the 1977 car model years, fuel economy increased 34 percent for new cars: from an average of 13.9 to 18.6 miles per gallon.[9] Consumption of gasoline fell below 1973 in 1974 and 1975, but was 5 percent above in 1976, a trend related to recession and recovery. The fuel economy standards for 1984 mean a 50 percent additional gain in efficiency, and an expected 13 percent decline in consumption of gasoline. Far more important than fuel economy, however, is emission control technology. The reduction in emissions below precontrol levels required by the interim standards of 1975, which have been attained, was 83 percent for HCs, 83 percent for CO, and 32 percent for NO_x.[10] The emission control technology used to reduce HC and CO emissions in 1968-72 model cars increased NO_x emissions. The addition of exhaust gas recirculation in model years 1973-74, and the advent of catalytic converters in 1975, counteracted this effect, so that in 1975 total NO_x emissions began to decline.[11] (There was an adverse effect of catalysts, not covered by auto emission standards: their increased emission of sulfates.) By 1977, motor vehicle emissions of NO_x were down 43 percent when compared with those of 1970, whereas HC and CO were down 80 percent. These are the pollutants mainly attributable to motor vehicles. It is not surprising, therefore, that national emissions of both HC and CO declined steadily from 1970 on. NO_x showed a slight upward trend until 1975, when it too began to decline. Very few areas have measurements of oxidants, usually represented by ozone, over this time period. However, Los Angeles, subject to California standards (which are more stringent than national standards), experienced a declining trend over a five-year period.[12] Particulates and SO_2, mainly attributable to industry, declined substantially over the same period.

Relevance of Regulations for Health

The policy of nondegradation can be justified on a health basis (cost aside) only if there is no minimum threshold—if any amount of air pollution, no matter how small, is a threat to health. But many studies have failed to find evidence of damage at low levels of pollution. A more plausible theoretical case can be made that there is no harmless level of pollution for carcinogens.

The expansion of the number of stationary sources subject to emission control can be justified in those areas where otherwise the level of pollution would be or become unacceptable on health grounds. But blanket emission controls for incinerators, petroleum storage tanks, and other facilities in clean and dirty areas alike makes no sense on a protection-of-health basis unless, again, one assumes a zero threshold, or some threshold well below national primary standards, for health damage.

Control of complex sources of pollution that attract concentrations of automobiles, resulting in excessive local levels of pollution, on the other hand can be justified on health grounds. Such controls are only applied in situations where excessive pollution is expected. The only controversy on health grounds is whether national primary standards are indicative of threats to health. To these standards, and associated emission standards, we now turn.

How Ambient Air Quality and Emission Standards Were Set

The National Air Pollution Control Administration states:

> The chemical and physical characteristics of the pollutants and the techniques available for measuring these characteristics must be considered, along with exposure time, relative humidity, and other conditions of the environment. The criteria must consider the contribution of all such variables to the effects of air pollution on human health, agriculture, materials, visibility, and climate. Further, the individual characteristics of the receptor must be taken into account.[13]

Decisions on reasonable air quality standards were relegated to the "prudent judgment of experts in various disciplines," who used as a basis the criteria documents published by the National Air Pollution Control Administration.[14]

Hazards to human health were estimated by several means. Working people, athletes, the general population, schoolchildren, and particularly persons suffering from chronic bronchitis, emphysema, or asthma were questioned or observed in action during times of high and low levels of one or more pollutants. The relation of auto accidents to high and low levels of each of several pollutants and the extra risk of dying from a myocardial infarction during periods of high concentration of several pollutants were examined. The incidence rate of respiratory ailments was studied in a general population and in schoolchildren, in areas with high and low concentrations of one or more gaseous pollutants.[15]

The primary ambient air quality standards were established by making the most conservative estimate of possible health damage for each pollutant, then adding a generous margin of safety. The estimate of the minimum critical concentration was based on those studies available at the time which showed possible damage to health at the lowest concentration of each pollutant. The EPA is said to have interpreted the Clean Air Act to require a "worst case" philosophy in estimating desired emission levels.[16] A "worst case" approach would mean using the highest recorded values of existing pollutant concentrations and the highest observed natural background level and assuming that they are maintained indefinitely, and using the lowest estimates for harmful concentrations. This procedure would require maximum reduction of emissions from man-made sources of pollution.

The target reduction of motor vehicle emissions was intended to allow attainment by 1980 in every city of the primary standards for the pollutants which are largely the products of motor vehicle operations: CO, NO_2, hydrocarbons, and oxidants. The 1975 auto emission standards made the most conservative assumptions about ambient air pollution (taking Chicago as the standard for CO and SO_2, and Los Angeles for NO_2, hydrocarbons, and oxidants) and the maximum estimate about the rate of growth in the number of motor vehicles. Emission standards then were based on a set of three worst-case assumptions. (They took into account natural background levels and assumed that emissions of CO, NO_x, and hydrocarbons from stationary sources could be reduced by a "like percent.")[17]

Two additional assumptions had to be made for CO and oxidants. The critical concentration of CO in health impact studies referred to carboxyhemoglobin in the blood (3 percent). This had to be correlated with CO concentration in the ambient atmosphere, and a conservative assumption was used here, too: much lower ambient concentrations of CO were required than were used in health impact studies to attain the critical carboxyhemoglobin levels in the blood. Oxidants are the product of complex photochemical reactions involving hydrocarbons and oxides of nitrogen, facilitated by trace catalysts. The relation between hydrocarbon emissions and oxidant levels again was based on the most conservative assumption, a linear relation.

All assumptions listed so far have the same bias: toward overstatement of health damage, or pollutant concentration, and therefore toward unduly strict air quality standards. An additional assumption also is biased in the same direction: that air pollution concentration is proportional to the rate of emission of pollutants. This would only be true in a contained space, and can only be approximated in the worst cases: temperature inversion, no wind, and basinlike topography. A further assumption is that all the NO_x, CO, hydrocarbons, and resultant oxidants are derived from operation of motor vehicles. To the extent that this is not true (it may be approximated in the worst cases), motor vehicle emission rates have less than proportionate effects on ambient air pollution concentrations. The motor vehicle emission standard required to attain the primary air quality standard depends on the importance of other sources of these pollutants, and the extent to which their emissions are reduced. Thus, the bias in this assumption could go either way.

Another assumption is that the effects of the different pollutants are independent. There is some evidence that they interact, and that the combined effect of some pollutants, for instance SO_2, NO_x and particulates, is greater than their separate effects. In this regard, emission controls and air quality standards may not be strict enough.

The deadline of 1975 for meeting the proposed standards was chosen by the Committee on Public Works of the U.S. Senate on the basis of information received and hearings held over the years (1964, 1965, 1967, and 1970). The committee was aware that the required technology did not exist, but assumed that it would be developed within the period indicated. Repeated postponement in standards makes it clear that that assumption was wrong.

Questions have been raised about the adequacy of the evidence on which the standards have been based.[18] Evidence that the EPA's decisions, as well as decisions embodied in legislation, were recognized to be arbitrary–based on considerable ignorance–is provided by the changes that have occurred to date in standards and deadlines. Recognition of ignorance is also evident in the EPA's requests for research proposals for fiscal year 1973, well after standards were set, which called for quantification of health impacts of human exposure to sulfur oxides, NO_x, particulates, photochemical oxidants, CO, acid mists, fuel and fuel additive emissions, and so forth. Also called for were studies of pollutants acting singly or in combination with other pollutants or environmental factors, to provide information for development and revision of air pollutant criteria and standards.[19]

California, with very severe air pollution problems, established its own standards in advance of national ones, in the light of much of the same evidence available to the Congress and the EPA. Its "serious" level, roughly corresponding in definition to the national primary standard, in every case allows higher concentrations of pollutants than the national standard: 3 ppm maximum one hour for NO_2, versus 0.05 ppm annual average; 120 ppm maximum one hour for CO, versus 35 ppm; 5 ppm maximum one hour for SO_2, versus 0.5 ppm maximum three hour concentration. For oxidants it defines only an "adverse" level at

1 ppm one hour maximum, versus the secondary standard of 0.05 ppm three hour maximum. For ethylene the "adverse" standard is 0.5 ppm one hour maximum, versus 0.08 one hour maximum for hydrocarbons nationally.[20]

Setting standards, however, is not a simple difficulty of measurement, but a subtle problem of valuation. The facts are that vulnerability to air pollution varies widely among the population, and that the extent of air pollution varies widely by area. Emission standards set to achieve acceptable ambient air standards for the most vulnerable population groups in the most polluted areas have a dubious justification when applied to everyone everywhere. In each area, there is no practical way of requiring different standards for population groups with different rates of exposure or different vulnerability. Between areas, it is practicable, up to a point, to vary emission standards to achieve a nationwide ambient air standard.

The American population is mobile; many migrate from areas with low pollution to areas with high pollution or with different pollutant mix. Hence, it may be reasonable to set ambient air standards to avoid harm to the health of population groups that have not built up a tolerance to pollutants. A standard adequate for those living in unpolluted areas and for the very young, who have not built up a tolerance, however, is unnecessarily strict for most of the population.

Ambient air standards and emission standards were set on the basis of studies predominantly of the most vulnerable groups: young children, asthmatics, emphysematics, cardiac patients. Asking what the standards should be amounts to asking who should be protected. The stricter the standards, the fewer the individuals who need them to protect their health. The question is ultimately a moral one: How much are the many prepared to pay for assuring zero risk from air pollution to the health of the few? Everyone, of course, was once a newborn.

How Should Standards Be Set?

The primary objective in reducing air pollution is improvement of health. The optimal policy toward this objective depends on two relations: (a) the relation between changes in the level of air pollution and harmful effects on health; and (b) the relation between reduction in air pollution and the costs of reducing it.

If we could relate specific concentrations of particular pollutants to specific physiological consequences, quantitatively measurable (reduced effectiveness, illness, mortality), then we could calculate a trade-off between pollution abatement and health. But health (or the lack of it) is not measurable by any simple unit. Mortality may appear to be measured so; what is at issue, however, is not excess mortality during acute pollution episodes, but foreshortened life, which involves estimation of expected life in the absence of pollution. It is straining to convert days ill into foreshortened lives. Milder lapses from health, which cannot be expressed in bed-days, or even sick-leave days, also must be included in the impact measure. Perhaps, therefore, several trade-offs should be calculated: pollutant concentration and mortality; pollutant concentration and bed-days or sick-leave days; and a residual for lesser degrees of reduced health which have no satisfactory indicant: not sick-leave or bed-days, not doctor visits or medication.

The necessary next step, assuming the first is achieved, is to convert both health effects and costs of pollution abatement into a common unit; specifically, dollars. Then it is possible to stipulate the optimal level of pollution abatement as that where the incremental cost is just equal to the incremental value of the resulting improvements in health. This means quantifying the cost of premature death, of illness, of poor health. Because of large interregional differences in both costs of and benefits from abatement, a national standard equating aggregate incremental costs and health benefits will be too stringent for some regions and not stringent enough for others.

In brief, if health effects were known, pollution standards could be derived from them. Since, in fact, health impacts are not known, other ways must be devised to arrive at air quality standards.

In the presence of ignorance about health effects, what standards should be set? Most people would agree that the presumption of innocence until proven guilty is not applicable when the accused are chemicals in the air, and the crime is illness and premature death. They would hold that the burden of proof is on the polluter. Realistically, however, penalties are imposed on producers, whose costs are raised, and ultimately on consumers, whose prices are increased. So the issue appears to be one of imposing penalties on property when the consequences of its current uses for health and life are uncertain. This is a false dichotomy: property versus life and health. The penalties require consumers and producers to devote more of their life and energy to produce a quantum of goods. The choice is between lives constrained by a penalty on production and some lives constrained by a penalty on health.

Whatever our present evaluation, more information in the future will alter our judgment. New pollutants may be recognized as threats (as polyvinyl chloride has been only recently), their relative importance will be reassessed, and the relation between pollutant concentration and health and other effects may also be revised. This current ignorance, remediable in the future, argues for flexible policies which can be altered fairly readily in the light of better information, keeping in mind of course the costs associated with change and uncertainty in policies. Since our ignorance refers also to the rate and direction of technological progress in emission control, inflexibility as to air quality and emission standards, required techniques, and the timing of either is ultimately untenable.

If there were a constant relation between an absolute change in air pollution and the health costs associated with it, on one hand, and the costs of abatement, on the other, it would not matter where a given amount for pollution abatement was spent. The results in terms of better health would be the same. That is, technically and economically it would not matter; ethically, it would. A given amount can significantly improve the health of a minority, or marginally improve the health of nearly everyone. Our democratic tendency is to distribute the benefits widely. But does it mean equal improvements in health everywhere, or equal health (so far as air pollution is concerned) everywhere? Since areas vary widely in levels of pollution, the second, redistributive concept of equity would concentrate pollution abatement and health improvement in a limited number of the most polluted areas, neglecting most other areas.

It is assumed that the distributional objective is the equalization of the health effects of air pollution everywhere, rather than equal reductions everywhere. This implies nationally established area standards which will eventually approximate the actual conditions of the least polluted areas.

The questions to ask in selecting air quality standards and air quality policies are:

1. Is the total benefit worth the total cost (referring to an air quality standard, assuming optimal policies to attain it)?
2. Is the selected air quality standard the one offering the maximum net benefits, considering the variation of both benefits and costs with the standard selected?
3. Are there more cost-effective policies of achieving a given air quality standard?

Because of diminishing returns from available techniques, benefit-cost ratios for a 50 percent emission reduction will be much greater than for a 75 to 90 percent reduction. On the other hand, the maximum net benefit (benefit minus costs) will be for a standard higher than that yielding the highest benefit-cost ratio. Furthermore, higher benefit-cost ratios, and larger net benefits, are attainable for a given percentage reduction of concentration for some pollutants than for other pollutants. Thus, the emission-reduction mix should be considered in optimizing policies and standards.

Benefits can be defined as the difference between the damage occurring under unregulated conditions and the damage occurring under controlled conditions. To define the optimum level of pollution control, both damage costs and control costs must be known.

Under the Clean Air Act, the EPA must consider compliance costs in establishing standards of performance for new stationary sources. This approach clearly should have been applied also to motor vehicles, for which cost of compliance was not considered, and to air quality standards. Neither section 108 (which provides for issuance of air quality criteria and information on air pollution control techniques), section 109 (which directs the administrator of the EPA to promulgate air quality standards), nor section 110 (which requires states to adopt and submit implementation plans) provides for any consideration of costs versus benefits. There is rarely any justification for a policy that counts no costs. In fact, repeated postponements of deadlines for compliance reflect EPA and congressional recognition that costs were unacceptable. One ignores large costs only when the benefits are expected to be indefinitely large. Yet the damage done by stationary sources, for which costs counted from the start, is much greater than that done by motor vehicles, as well as much more evident and certain. Hence, the benefits of emission abatement must be greater for stationary sources than for motor vehicles. Inflexible deadlines for compliance, in the absence of the necessary technology, is another way of ignoring costs.

> Publication of proposed rule-making should be done only after the costs, benefits and side-effects of the intended regulation are understood to the extent possible by the parties concerned. Failure to carry out this investigation phase adequately may lead to unnecessary economic losses, loss of regulatory agency credibility, protracted formal processes, and litigation . . . . Establishing standards beyond the known state of the art on the theory that industry can do anything if enough pressure is put on it is not likely to result in wise governmental decision making or to provide the greatest net benefits to society.[21]

5 Benefits of Pollution Abatement

Clean Air in Relation to Other Goals

Clean air is not our only goal. Ideally, an assessment of the benefits of clean air would include its contribution to or interference with attainment of other goals. One cost of clean air is dirty water and a litter-strewn countryside. Incineration of combustible waste does wonders for litter, but pollutes the air. Dissolving pollutants and pouring them into our waterways is a partial alternative to polluting the air. Precipitation of particulates from the air is responsible for water acidity and damage to vegetation. Washing the air just transfers gaseous and solid airborne pollutants to land or water, where they may, or in some cases may not, become harmless. There are some trade-offs between environmental goals. The sacrifice of other goals should be subtracted to obtain the net benefits of clean air. The main conflict is between burning solid waste—incineration—or dumping it into rivers and seas. It is clean earth practices which are likely to be in conflict with cleaner air and cleaner water goals and policies, not vice versa.

But there are other, positive interdependencies. Water cannot remain pure in a polluted atmosphere; there is too much interaction between air and water. Nor can the land be clean if the air is dirty. Airborne pollutants mostly end up on land or sea, whether in their original form or as compounds formed by chemical reactions in the atmosphere. Water acidity is a very widespread consequence of air pollution in industrial areas. Reduction in air pollution will reduce the deposits of pollutants in soil and water. Thus clean air, clean water, and clean earth are typically complementary ends, not achieved at each other's expense. Naturally they are separate goals, best attained by separate techniques; cleaning one element is not necessarily an effective way of improving another. Nevertheless, part of the benefit of clean air should be its contribution toward cleaner land and water.

On the other hand, clean air and energy goals are in sharp conflict. The central aim of energy policy is to reduce United States dependence on imported petroleum products. The main strategy being considered and to some degree implemented is greater reliance on domestic coal as the alternative fuel. But coal is by all odds the most polluting energy source available, accounting for much of SO_x and particulate emissions. Its expanded use significantly raises the cost of attaining a given level of air quality or pollutant emission, and aggravates problems of watershed pollution near coal mines and of waste disposal for power plants. Benefits of emission reduction achieved by discouraging the use of coal, or by requiring the use of scarce and expensive low-sulphur coal, are clearly attained at the expense of the central strategy of our energy program.

Not all strategies are in conflict; the secondary approach of energy conservation complements clean air goals. But the basic conflict prevailing between environmental and energy goals will concern us throughout much of the remainder of this book.

Present Versus Future Benefits

Estimates of the benefits of air pollution abatement invariably take existing levels of air pollution as a benchmark. The benefits of eliminating all air pollution are equal to the damages of the extant level of pollution. The benefits of a given absolute amount of abatement, or of a given percentage reduction in pollution, will vary with the initial level of pollution. On the assumption that the air pollution damage function is exponential rather than linear, then the higher the initial level of pollution, the greater the benefit of a given abatement effort or amount or percent.

What we face today is a somewhat different situation: the threat that air pollution will increase if greater abatement efforts are not undertaken. This threat includes the long-run consequence of increasing population numbers, and particularly of increasing affluence. Per capita income rather than population alone accounts for most of the growth in air pollution. In addition to this long-term trend, the newly conceived effort to reduce American dependence on imported oil has led to plans to increase coal production and use, in some cases to compel reconversion of furnaces from oil to coal, and to require that new utilities and industrial power plants use coal. If one is to take such plans as a measure of the future level of pollution without abatement measures, then the benefits of abatement have begun to rise far more steeply than anticipated in the past. Existing studies of benefit all understate to the extent that they were based on existing levels of pollution, rather than on the levels that would prevail should energy independence plans proposed by President Carter be implemented.

Levels of air pollution and corresponding emission standards that are acceptable today may not prove acceptable in the future. What is acceptable is typically related to what is attainable, which will change with modifications in environmental conditions and improvements in pollution abatement technology. What is acceptable and how the benefits of abatement may be perceived depend also on awareness, which has grown and will no doubt grow further with publicity about air pollution hazards and with attitudes that have become less permissive toward perceived risks. Standards will change. The air quality standards legislated in 1970 will not be fully implemented until 1987, if then. The lead time for pollution abatement policies is long enough to incorporate changes in circumstances, technology, and attitudes. To the extent feasible, therefore, standards should be based on projections and/or forecasts of benefits (and costs).

Benefits to Health of Reducing Air Pollution

The procedure for estimating the benefits of reducing air pollution is protracted. First, emission reduction is measured, which then results in lessened air pollution, but not in any simple ratio. Reduced air pollution cuts down on pollution exposure, but not proportionately. Then, reduced pollution exposure is associated with reduced mortality and morbidity—an association demonstrated to be complex and uncertain. Finally, a value is placed on the decline in mortality and morbidity. Thus, the relation between emission standards and health benefits is connected by many loose links.

National emission standards, which are designed to lower ambient air pollution in the worst city below national air quality standards, reduce emissions and improve ambient air quality everywhere, even in areas where air quality is good. The benefit-cost calculations that may justify emission standards in the worst city would not necessarily justify them anywhere else, and certainly not in many areas. The costs of some emission reduction policies, e.g., automobile exhaust standards, are roughly the same everywhere; they do not depend on the level of air pollution. On the other hand, the benefit is a function of the level of air pollution.

Available evidence suggests that increasing pollution has progressive rather than proportional effects on health. This means that equal absolute reductions in pollution do not have equal effects on health. Below the threshold at which health damage occurs, the benefits as far as we know are zero. The distribution of benefits is an equity objective whose implementation is also influenced by the pollution-health relation at pollution levels below the air quality standard. If it is assumed that the standard is the threshold for damage to health, all reduction in pollution should be in areas that do not meet the standard.

For some pollutants, such as carbon monoxide, critical levels of concentration affect health. For others, such as lead, cumulative exposure is the culprit. It should be noted that the primary standards for different pollutants are expressed in terms of concentration for different time periods (see table 4-1). Optimal area policy depends, therefore, on the composition of the area's air pollution. But every area will benefit more from reducing pollution in peak rather than in off-peak hours, days, or seasons. Measurements for a number of metropolitan areas show that the annual average level of most pollutants does not vary a great deal between cities, but the shorter the time period—maximum seven-day average, twenty-four hour average—the greater the difference between cities.[1]

Ambient air standards may be said to place zero value on variations in pollution which do not exceed the standards. On the other hand, the policy of environmental nondegradation implies no minimum threshold to health damage. Can areas that fall well within the clean air standards set by the EPA allow their air quality to deteriorate? If there is a linear relation between pollution and health throughout the range of pollution observed in this country, then regulation of areas with clean air would be beneficial to health. But if health is damaged more than in proportion to increasing concentration of pollutants, industries should be redistributed to less polluted areas (considering the benefit side only).

Techniques that eliminate one type of air pollutant may increase the emission of another type. Different kinds of benefits (reductions of different kinds of pollutants in varying degrees) must be rendered comparable to calculate trade-offs. For instance, elimination of lead from fuels will reduce the efficiency of engines, which will consume more gasoline, and therefore generate more of all other pollutants. Is this a good bargain, or not?

If measured air pollution is to be a reliable index of exposure to pollution, even on an aggregate level, account must be taken of the difference between outdoor and indoor pollution and of the time spent in each environment. There are large regional differences associated with climate. Most people spend much of their lives indoors, where the level of pollution is typically far below, and of different composition from, that of outdoors. If total cumulative exposure is what matters for health, then the difference in pollution indoors and outdoors and the time spent in each environment should be taken into account. If, on the other hand, there is a minimum threshold of harmful effects, which is typically higher than indoor pollution, then only outdoor pollution counts.

The final problem is in placing a value on estimated reduction in mortality and morbidity. One approach is to value mortality in terms of expected future earnings foregone. This is used by Lave and Seskin, who regard it as a gross underestimate. This approach is highly sensitive to the person's age at death and to the extent to which life is foreshortened by pollution. It can be modified by subtracting the consumption of the individual affected. It implies that the life of an individual who has no expected future earnings has no value. Schelling's suggested measure is the individual's willingness to pay to reduce the probability of death, plus the value society places on the individual.[2] Unfortunately, the pertinent data indicate an extremely wide range in valuation of human life. Decisions are made either in ignorance of probabilities, or with other objectives in mind as well. The unwillingness of most people to fasten seat belts implies a very low valuation of their own lives, whereas their willingness to buy flight insurance implies a very high valuation.

Assessing benefits implies a standard of measurement. Benefits imply satisfactions, which in turn imply preferences and expectations. Decisions on the level of environmental quality, however, are not made by individuals or firms. They are made by governments which express the collective judgments of some of their constituents. Willingness to pay could be proposed as a measure of benefit for pollution abatement, as it is for most goods and services. But the latter are bought and sold in the market and have determinable prices, whereas there is no such market for pollution abatement. Willingness to pay may be impossible to assess, and is inadequate as a measure of the benefits of air pollution abatement. The basic reason for this is the inevitably inadequate awareness of the harmful effects of pollution and incorrect valuation of the benefits from abatement. Just as there is no close relation between polluter and victim, there is little perceived relation between the costs of abatement and the benefits received. Air pollution abatement is a classic example of a "public good" whose consumption cannot be limited to specific groups. Cleaner air benefits everyone in an area, whether or not they are aware of the damage of pollution and the value of abatement, whether or not they are willing to pay for it. Under these conditions, even the informed individual has no incentive to reveal his true valuation of expected benefits or his willingness to pay. For both reasons—ignorance and the impossibility of limiting benefits to those who pay the costs of abatement—the willingness-to-pay criterion must be rejected and objective evaluations of benefits sought out instead. (A different, and misconceived, reason for rejecting willingness to pay is its assumption that victims rather than emitters ought to bear the burden of abatement. In fact, costs imposed on polluters are partly if not entirely shifted to the purchasers of their products, who have also benefited from free use of the environment for waste disposal. The users or beneficiaries are often also the victims.)

Two studies often cited are those of Ronald G. Ridker (1967)[3] and of Lave and Seskin (1970)[4]. Ridker estimated the cost of air pollution by sulfates and particulates only, and their impact on respiratory diseases and cancer of the respiratory system. The costs included were loss of output, premature burial, and treatment. The total for 1958 (using a 5 percent discount rate) was just under $2 billion.

Lave and Seskin estimated the benefit of a 50 percent reduction in air pollution in major urban areas at $2.1 billion a year. They regarded this as a gross underestimate because, first, it was limited to the effects of particulates and sulfates, and second, its valuation of human life and health was limited to the value of foregone earnings and the direct costs of disease. In addition to respiratory diseases and lung cancer, they included mortality from all other cancers and cardiovascular mortality and morbidity.

The estimates are not quite comparable, since Lave and Seskin employed a narrower definition of costs but applied it to a wider range of diseases. Lave and Seskin's estimates are clearly much higher, however, since they refer to a 50 percent reduction whereas Ridker implies complete elimination of air pollution. Both studies assume a linear relation between air pollution and health.

Considering the uncertainties in measuring air pollution and its impact on human health, not much credence can be placed on the magnitudes indicated in these or other estimates. Health damage is the main, but not the only, cost of pollution or justification for pollution abatement.

Larry B. Barrett and Thomas E. Waddell of the EPA estimated total national health costs of air pollution for 1968 at $6.06 billion. With no pollution control, these costs would rise to $23 billion in 1977.[5] These costs were for only two pollutants, SO_x and particulates. By emission source, over half the cost—$3.281 billion—was contributed by stationary sources of fuel combustion; $1.458 billion by industrial processes; $197 million by transportation; $119 million by solid waste disposal, and $1.005 billion, miscellaneous. The RECAT (Regulatory Effects on the Cost of Automotive Transportation) Committee extended this estimate of total health damage by including the effects of CO, hydrocarbons and NO_x. Its estimated damage ranges from $8.886 billion, using Babcock's severity factors to weight these added pollutants relative to SO_x and particulates, to $15.169 billion, using Caretto-Sawyer factors to weight them.[6] Motor vehicles account for not quite one quarter of the total, using the Babcock factors, or slightly over one third, using the Caretto-Sawyer factors.

Lave and Seskin, in their latest and most comprehensive study on the subject, greatly increase their earlier estimates, arriving at totals similar to those of Barrett and Waddell.[7] They conclude that "there is little evidence indicating an important association between automobile emissions and ill health on a nationwide basis." Even in Los Angeles there is no unequivocal relation between smog and daily deaths. Their estimates of health benefits are accordingly limited to the effects of reducing sulfate and particulate pollution. Their estimate is that a 58 percent reduction in particulates and 88 percent in sulfur oxides relative to the 1970 control levels will reduce mortality on the order of 7 percent. They assume that the effect on morbidity will be of the same order of magnitude. The next problem is placing a value on this estimated reduction in mortality and morbidity.

Although Lave and Seskin reject the human capital (or foregone earnings) approach as a conceptual basis for estimating the benefit of reducing pollution, they are compelled to rely on foregone earnings plus direct medical expenses of a typical United States resident because of the unavailability of data appropriate for alternative concepts of benefit. They arrive at a total of $16.1 billion in 1979 (expressed in 1973 dollars) for the health gains from abatement of particulate and sulfate pollution. This is the estimated value of a 7 percent reduction in the total economic cost of mortality and morbidity.

These last benefit estimates are more accurately described as estimates of direct expenditures and foregone income associated with the health and mortality effects of air pollution. As such, they are very much a lower boundary. They do not fully reflect the willingness of individuals to pay in order to avoid these consequences, as well as other consequences not involving expenditures. They are limited to the estimated impacts of the pollutants for which national ambient air quality standards are in force, or only to some of them. These estimates do not include, for instance, the effect of acid rain, which affects the health of populations far beyond the polluted cities and industrial centers where the SO_x and NO_x that are responsible are generated. We have become aware of this effect only recently, and not much is known yet of its consequences for health. There are no estimates of the health effect of lead in the air, for which a 1982 standard has been set, or its associated costs and expenditures. (Since most lead is absorbed through food and water, it would be difficult to arrive at a separate estimate for airborne lead.) Nor are there any estimates for asbestos, which appears to be on the way toward a national air quality standard of its own. Numerous air pollutants localized in concentration and typically associated with specific industrial activities are likewise excluded. So are the aeroallergens, whose effects are widespread, large, and well-known, but ignored in national legislation. In 1964 it was estimated that there were eight million sufferers from hay fever, who lost twenty-five million days from work and spent at least $25 million on medication. "The aero-allergens probably impose a greater cost in impairment of health than can be ascribed to any of the atmospheric pollutants for which control measures are now being developed."[8]

A further limitation of estimates comparing air quality standards in two different years is that they fail to calibrate for levels and changes in levels of pollution which differ from the standard, and for the influence of regional differences in pollution and exposure on aggregate health effects.

Uncertainty and Valuation of Benefits

In the absence of reliable estimates of the health costs of air pollution and of pollution abatement, or even in their presence, standards will be chosen and enforced in a democracy on the basis of bargaining between divergent interests. Precise knowledge would limit the range of possible bargains, although it would not eliminate dispute, since there are both gainers and losers among individuals, industries, and localities. But the present situation of limited knowledge, much uncertainty, and widespread ignorance encourages extreme positions. The goal of zero pollution is an atavistic wish for a pristine nature that never existed, a pre-Industrial Revolution society with no pollution. Nature is dirty. The opposite goal, of economic growth and never mind the by-products or casualties, is equally extreme, a future-orientation that values the welfare of the current population at zero.

The dilemma posed for policy is illustrated by the implicit claim of infinite value for the Furbish lousewort, a humble little plant now enjoying its moment in history. As the London *Economist* has pointed out, stopping progress to save the lousewort is an attempt to outlaw the workings of Darwinian evolution. It amounts to an inflexible commitment to the status quo, to the point that opponents of any change now seek to identify some minute speciation in plant or animal life in an affected area and certify it as an endangered species, endowing it with infinite value. Placing the survival of some obscure subspecies of plant or fish on the same standard as the welfare or even the lives of human beings reflects what Kenneth Boulding calls the "heroic ethic" that never counts the costs or the benefits. Of course, there are illustrations at the other extreme: the "highwaymen" with their addiction to concrete and the "growth-maniacs" are the counterparts of the "econuts." Extremism, a myopic perspective, is the hazard of those with singular goals in a pluralistic world.

Whereas some of the costs of air pollution, such as dirt and damage to materials, are certain and measurable, in principle, the health costs to individuals may always involve probabilities. In our current state of knowledge, we must deal not only with risks (calculable probabilities, which approximate certainty for large numbers under specified conditions), but also with uncertainty, which is unquantifiable. Pollution abatement reduces the subjective risks and uncertainties so far as the individual is concerned, but does not eliminate all possibility of pollution-related illness. What, then, is an acceptable level of risk and uncertainty? What is the trade-off between ever lower risks and uncertainties, and ever higher costs of pollution abatement? The answer appears to depend on the type of eventuality by which one is threatened. Individuals (and their government officials) are prepared to go to near-absurd lengths to avoid minimal exposure to carcinogens, but are more willing to live with high risk and uncertainty when the consequences may be bronchitis or asthma. Objective measures of the benefits of pollution abatement, even if accurate and well known, will not adequately reflect the concerns and anxieties of individuals and their representatives.

A recent case in point was the public alarm over circumstantial evidence, based on animal tests, that anyone regularly imbibing some eight hundred bottles of sugar-free soft drink daily would ingest enough cyclamate to risk developing bladder cancer eventually, if somehow he managed to avert drowning in soda water first. (The greater hazard of ingesting the equivalent amount of sugar instead was overlooked.) On the other hand, much less alarm has been evident over the difference regarding statistically predictable death and disability between a 50 and a 55 mph speed limit. In sum, attitudes toward uncertainty and the kind of tunnel vision that is blind to alternatives and consequences are elements that should not be ignored in the political assessment of benefits and costs.

Other Benefits

It should be recalled that whereas primary standards are intended to protect the health of the public, secondary standards are to protect the public welfare from any adverse effects associated with air pollution, including effects on soil, water, vegetation, materials, property, animals, wildlife, weather and climate, as well as economic values and personal comfort and well-being. Cleaner air has many benefits other than better health, most of which, however, are even more difficult to quantify. Most often mentioned is the gain in property values, or, more correctly, the avoidance of the loss of property values associated with air pollution. Studies comparing market values and rentals of comparable buildings in areas differing only in air pollution have arrived at some rough measure of this value. The problem with loss in value of property is that it reflects a multitude of causes, and involves much double counting. For example, to the extent that people are aware of the health hazards of heavily polluted areas, the discount at which they buy and sell property in such areas includes their capitalized value of health consequences. To some extent, air pollution is associated with traffic congestion, which may also be found objectionable in itself: delays in access, noise, reducing the amenities of the area.

There are costs of damage to materials associated with air pollution. Exposed surfaces become dirty sooner; painting, washing, and laundering must be done more often; the life of fabrics, rubber, and other materials is shortened. There are also aesthetic costs of dirtiness, of bad smell, of poor visibility.

The benefits of air pollution abatement are joint. Benefits to health are inextricably associated with aesthetic benefits, economies in the life of materials and in cleaning bills, frequently also with reduced congestion costs. The jointness of benefits, however, does not extend over the entire range of pollution abatement. Some pollutants, for instance carbon monoxide, with potential damage to health, are colorless and odorless and have no aesthetic effects. The same is true of some carcinogens. Levels of pollution that do not affect materials may damage health, or vice versa. Nevertheless there are joint costs of pollution and joint benefits of pollution abatement, so that it is unrealistic to evaluate policies strictly on the basis of their effect on health without regard to their effects on other objectives in the agenda of society.

This discussion is necessarily limited in scope to health. The fact that a policy may not be justified on health grounds alone does not mean that when other benefits are considered it will not be a desirable policy. On the other hand, if a policy is justified on health grounds, it is more than justified when other benefits are considered. However, it is not possible to multiply the health benefits by a constant in order to arrive at total benefits. In some cases the constant would be one; in others, a much larger number. Sulfates have serious effects on materials and aesthetics as well as on health. Nitrogen oxides have little effect on materials and much more on aesthetics. Carbon monoxide affects neither. Particulates raise cleaning bills, are aesthetically offensive, but may have little effect on materials unless they are chemically reactive.

This brief discussion of benefits of air pollution abatement other than to health is motivated by two considerations. The first is the jointness of benefits: abatement necessarily reduces all kinds of damage, not just health damage. The second is the magnitude of damage (and of abatement benefits) to other things than health. Other kinds of damage appear to be of the same order of magnitude as health damage—in fact, somewhat larger than health damage. Policy on air pollution, therefore, cannot be based exclusively on considerations of health.

Land Values

Perhaps the most comprehensive approach toward estimating the costs of air pollution is its effect on property values. Particular attention has been devoted to residential property values, although there is no reason to believe that other types of property would be unaffected. The most widely cited estimate of the effect of air pollution on residential land value is that of Ridker and Henning for St. Louis.[9] They found that values in the highest-pollution areas in 1960 were up to $1,715 less than in lowest-pollution areas, holding constant other variables affecting land values. Air pollution explained 1.2 percent of the variance in median land values. The pollution index used was sulfation levels. Anderson and Crocker presented new data, including SO_x and suspended particulates, on St. Louis, Kansas City, and Washington, D.C.[10] They found a mean capitalized loss of $300 to $700 per residential unit, and a decline of 1 to 2 percent in offer prices for every 10 percent increase in composite air pollution.

A number of studies have examined the difference in values of otherwise comparable properties as a function of their proximity to major traffic arteries.[11] Although proximity to a high-volume traffic artery is unquestionably associated with differential air pollution, it is also associated with higher noise levels and neighborhood traffic congestion, and maybe with better access, so that it is not possible to attribute differences in values to the differences in air pollution alone.

These figures are highly speculative, and generalizing from them is questionable. The use of property values for estimating the benefits of marginal improvements in air quality can only be extended to calculating the value of nonmarginal improvements on the assumption that household valuation is independent of the level of pollution (and of household income).[12] But the biggest doubt concerns the attribution of the observed difference in property values to air pollution. These differences can only refer to buyer and seller perceptions, not to actual damages. Even with regard to perceptions, it is not the relative decline in or lower value of property in polluted areas that measures damages. We live in an imperfect world with wide locational choices. An increase in pollution that affects an area's property values adversely affects property values of alternative locations positively, or relatively so. It is the net effect on property values in the entire area within which people make residential location choices that is the closest measure of the damage of an increment (or benefit of a decrement) in air pollution.

As to the damage of existing unchanged levels of pollution, comparison must be made between rather than within cities, and introduces too many other factors that cannot be controlled. To compare values in cities with values in areas with near-zero pollution, we would have to go far away from concentrations of population and economic activity; the differences in land value would be attributable predominantly to other factors which could not be isolated from pollution effects. Pollution, employment, economic opportunity, social and cultural variety are joint products. It is perfectly possible to find samples of cities in which there is a positive association of pollution with land values without implying that people prefer pollution. In fact, there is a positive association between city size and both land values and air pollution. The practical difficulties of isolating the effects of air pollution abatement on property values are overwhelming.

The land value studies assume that the same kind of people live in more polluted and in less polluted residential areas in the same SMSA. But the Tiebout hypothesis asserts that people sort themselves out by suburb and neighborhood according to their tastes and priorities as well as income. The possibility that people who live in clean and in polluted areas have different preferences suggests that property value findings understate the effects of air pollution. We cannot assume that there are no differences between people in tolerance of pollution, or in preference for other attributes of neighborhoods that may be associated with pollution. Ridker's exhaustive efforts to find behavioral counterparts of differences in pollution, such as differences in frequency and intensity of cleaning and maintenance activities, household expenditures on cleaning supplies, etc., ended in failure.[13] If people care equally, there is no firm behavioral evidence, on either an interurban or an intraurban basis.

Smith and Deyak fail to find any significant effects of differences in air pollution on property values in their eighty-five-city sample, but find differences associated with local public services and property taxes.[14] Others have criticized the property value findings for lack of adequate homogeneity of the samples and for failure to consider other variables that might explain the observed differences in residential property values. The findings, moreover, are claimed to assume immobility of functions and activities in the areas examined. They fail to control for expectations regarding the permanence of new sources of air pollution and prospects of imposing emission control requirements.[15]

Part of the problem is that what is being measured is not actual damages, but people's perceptions of the undesirability of air pollution. One might expect that the growing publicity since 1970 about air pollution and its harmful effects on health has magnified perceived damages, accentuating concern and increasing willingness to pay for clean air, and that behavioral consequences, including property value differentials, should be much greater now than in 1960 (which is the date of information for the Ridker and Henning and the Anderson and Crocker studies). But the near impossibility of disentangling one of many dimensions affecting property values in what is a highly differentiated and complex market remains. That some damage occurs, that there is psychic dissatisfaction, there is no doubt. But none of the quantitative estimates has much credibility, any more than do the studies that find no statistically significant effects of air pollution.

Damage to Materials

Damages to materials, in a perfectly informed world, would be fully reflected (together with other harmful effects) in property values. Such damages can be estimated directly, whether or not they are known to owners and incorporated in their behavior. There are case studies and laboratory studies of the effect of various pollutants—SO_2, NO_x, ozone, or general air pollution—on rubber, leather, dyes, textiles, metals, and numerous other materials, and on whether pollutants shorten their lives and/or justify greater expenditures on maintenance. Damage includes physical and chemical deterioration as well as soiling of materials. In addition to damage, there are costs incurred to avoid or minimize damage. Care must be taken here to avoid double counting; if damage is measured in terms of more frequent maintenance, then maintenance expenditures cannot be added to it.

Laboratory studies have to be related to real conditions; studies in limited areas have to be generalized to the entire nation. Damage does depend on both concentration of the pollutant and duration of exposure. There are synergistic effects between pollutants. Humidity increases the damage of most pollutants. Among the more specific studies, Mueller and Stickney estimate the annual damage to rubber products by ozone at $226 million; Salvin estimates the damage to textile fibers and dyes by air pollution at $206 million a year; Spence and Haynie estimate the damage to paints by particulates and sulfur oxides at $704 million a year; Fink, Buttner, and Boyd estimate the damage done by sulfur oxides and other pollutants to susceptible metal products at $1.45 billion. A more global estimate by Salmon puts the total estimated damage at $3.8 billion a year.[16]

Damage to Vegetation

Although damage to vegetation in urban areas may be included in the property value effects of pollution in a perfectly informed urban real-estate market, the bulk of the damage to vegetation refers to the agricultural sector, including forests and associated outdoor recreational activities. The main damage to vegetation is the loss of agricultural output, both in yield and in quality. There is substantial verified damage to trees as well, affecting the yield of forest products.

There is abundant evidence of damage including dollar estimates, to various crops in various areas.[17] Most of the damage is caused by oxidants, PAN and ozone in particular, and by SO_2. But the evidence falls short of providing a total national estimate of damage, especially of damage from lower levels of pollution and in mild form (such as reduced yields), rather than clear pathological symptoms in plants. In addition, there are aesthetic losses associated with damage to ornamental vegetation and to urban trees and plants.

Man can avoid or reduce much damage by introducing crops or tree species that are relatively immune to prevalent air pollutants and by developing pollution-resistant varieties. In this case the damage may be the sum of the cost of such replacements plus any diminution in the yields or quality of products of the new varieties, plus any change in the costs of cultivation and caring for the plants. As in the case of damage to materials, some studies are based on laboratory experiments that did not precisely replicate the environment; others are limited to specific areas or even specific episodes of high pollution, and cannot be safely generalized to the entire nation or to the year-in-and-year-out consequences of air pollution. There is the further problem that most of the damage to plants can also be caused by factors other than man-made pollution: weather, natural pollution, microorganisms, and parasites. The damage caused by these other causes may be aggravated when plants have been weakened by man-made pollutants, whether airborne or waterborne. Beyond identifiable damage, there are always variations in crop yields, tree growth rates, and product quality, which are extremely difficult to allocate among the several contributory factors.

Benedict, Miller, and Olson attempt the most comprehensive but lowest estimate for the entire nation, $132 million damage to crops and ornamentals for 1971.[18] The National Academy of Sciences in 1966 estimated damage to agriculture alone at $325 million. These estimates are incomplete. They consider only the loss of output, replacement costs for plants, and the like. They neglect the role of vegetation in the ecology. Plants remove vast amounts of CO and other pollutants from the air, and acquire from soil and air vast amounts of nitrogen. Damage to plants by some pollutants reduces the plants' ability to remove other pollutants from the air. Reduced vegetation coverage and reduced retention of soil nutrients mean greater erosion, greater water pollution downstream, and greater requirements for nitrogen and other fertilizers. These consequences have not been quantified for the nation, although there are indicative case studies. For example, a hectare of pastureland near San Bernardino, California, can absorb 440 kg of CO per year. Westman estimated the cost of erosion due to air pollution damage of trees in a 4,000 hectare area of the San Bernardino National Forest at $27 million per year. Since CO retards nitrogen fixation by bacteria in legumes, a concentration of CO of fifty parts per million would cost $5.50 per hectare per year to replace the lost nitrogen fertilizer. Ozone, at levels of concentration exceeded on most days in the Los Angeles basin, inhibits the activity of nitrogen-fixing bacteria by 40 percent.[19]

Other

Other kinds of damage include damage to animals, particularly livestock; damage to outdoor recreational facilities and activities; aesthetic damages; and anxiety costs. The last two are probably unquantifiable; certainly they have not been quantified. Suggestions that observation of behavior (such as change of residence in response to aesthetic or anxiety considerations) may give some indication are not too hopeful. People move for many complex reasons, and the reasons they give are often not closely related to their main motivations.

The value of livestock is directly affected by air pollution and indirectly by the effects of air pollution on vegetation. As in the case of vegetation, the factors influencing the value of livestock and costs of its production are numerous and interactive, so that the contribution of man-made air pollution may not be possible to isolate.

Avoidance Costs

The damage caused by air pollution would be greater than it is if people did not take measures to avoid it. Avoidance costs should be added to pollution abatement costs to determine the total costs of reducing the damage done by air pollution. Abatement costs are closely related to initial levels of pollution, whereas avoidance costs are not. Since avoidance costs do not affect pollution, but only exposure or sensitivity to pollution, they are included not among pollution abatement costs but instead as part of the potential benefit from pollution abatement. Avoidance costs may include relocation of people, reduction of outdoor activities, or the use of protective devices such as filters and masks. With regard to materials, they include preventive measures such as protective coatings and maintenance, and substitution of new materials which may be more expensive and less satisfactory. It is very difficult to distinguish maintenance measures which are the consequence of air pollution damage from those intended to reduce such damage. Avoidance costs with regard to vegetation may include displacement of sensitive crops, the substitution of pollution-resistant trees and plants, and development of resistant varieties.

As people and governments become more aware of the potential damage of air pollution, they incur larger avoidance costs for a given level. Lower property values in polluted areas is in part an expression of avoidance costs. There is no estimate of health damage avoidance costs. Some measures, such as relocation, avoid health as well as other kinds of damage. But most avoidance measures, unlike abatement measures, refer to specific types of damage and hence to specific potential benefits. Although some measures are separable and quantifiable, in general avoidance costs are difficult to separate from diverse other costs and benefits associated with particular decisions. This is especially true with regard to the effect of air pollution on property values.

6 Costs of Pollution Abatement

The costs of air pollution control are joint costs of obtaining diverse benefits: improved health; better aesthetic quality of the environment; less rapid deterioration of materials; less frequent need for cleaning, painting, etc.; reduced congestion; healthy vegetation. Since health, aesthetics, cleanliness, and materials preservation (including the health of vegetation) are differently affected by different pollutants, and differently sensitive to varying concentrations of pollutants, one cannot allocate a constant share of costs of pollution abatement to health benefits irrespective of the type of level of pollution. To attribute all costs to health improvement, on the other hand, clearly overstates the true cost of health benefits. The justification for primary air quality standards, however, is expressed solely in terms of health; only secondary standards are concerned with other benefits. Hence the costs of meeting primary standards are properly allocable to health alone, with spillover benefits to other goals, whereas additional costs for meeting secondary standards are truly joint costs.

The energy problem and the Carter administration's proposed responses raise the cost of pollution abatement by increasing the amount of pollution to be treated. This increase is the consequence of emphasis on greater utilization of domestic coal to reduce reliance on imported oil. The main impact will be on electric power plants. Some policies, however, contribute both to clean air and to energy goals. Requirements for improved gasoline mileage in future model years and for reduced weight of automobiles are the principal example. The costs of such policies might be considered joint costs. It is clear, though, that the dominant, if not the sole, motivation in implementing these policies is energy conservation; other methods are being stressed to reduce exhaust emissions. Hence, it is appropriate to allocate the costs of such policies to the energy goals, and regard the incidental effects on air pollution as unintended but nonetheless welcome spillovers.

Legislation and regulation at the federal level are not the only controls on air pollution available. Many communities have long been concerned with air pollution, and restricted location of polluting activities or amount of allowable emissions long before 1970. Thus, although polluting industries did have wide options of locating, without constraints, many sites were only open with conditions on industrial levels of emission. Not all the expenses to reduce pollution can be attributed to federal legislation. Even in the absence of local ordinances, it was always possible to bring action for damages against local industries, constraining their freedom to pollute. Thus, the estimated cost of air pollution control should not be regarded as attributable entirely to recent federal legislation, in particular to the Clean Air Act Amendments.

Among the costs of air pollution control are: first, the costs of monitoring air quality; second, the costs of enforcing air quality standards and regulations—surveillance and inspection of pollution-emitting sources and prosecution of violators; and third, by far the most important, the costs of compliance with air quality and emission standards.

Monitoring and Enforcement Costs

The expenditures of the EPA and of state and local government monitoring, inspecting, and enforcement activities must be included among the costs. Some of these costs cannot be allocated to reduction of air pollution, however. They are joint costs of a variety of pollution-reducing activities, or even joint costs of government activities other than improvement of the quality of the environment. There are also joint benefits; reduction in air pollution improves human health, but it also reduces damage to plants and animals, lowers cleaning costs, increases the life of materials, improves visibility, reduces noxious odors, results in more clear and sunny days, and generally enhances the "quality of life."

Shortly after passage of the Clean Air Act Amendments in 1970, the costs of air pollution control programs (federal, state, and local government combined) were estimated to rise from $208 million in fiscal year 1972 to $275 million in 1975.[1] Actual federal costs in 1972 were $265 million, over half ($136.5 million) of which went for research and development ($207 million, plus $96 million by state and local governments, according to a Bureau of Census estimate).[2] Federal outlays totaled $274.3 million in 1975. An additional $82.5 million which could not be allocated between air pollution and other media was spent in that year.[3] Actual outlays exclude monitoring, surveillance, and enforcement by state and local authorities (although they include some federal grants for these purposes), and they also exclude the large but unknown costs of preparing environmental impact statements.

The main transaction cost is undoubtedly neither government monitoring and enforcement nor the costs of preparing environmental impact statements, although the latter have proved sufficient to generate a new industry on their own. They are the delays, often in years, in construction and production that result from a protracted review and approval process, which increases costs and, by postponing production, delays and reduces benefits, even in cases where a project is approved. At this time these costs have not been calculated. Some are incalculable, inasmuch as their effect is to deter investment and production.

However, there is evidence that monitoring and enforcement activities are inadequate, and that noncompliance is high. Assuming standards are optimal, then expenditures should be increased to the point where the benefit from greater compliance is equal to the added cost of inspection and enforcement. There is no information on this trade-off.

Everybody wants clean air. There is some doubt, however, about whether everybody wants to pay the costs of having clean air. There are reports of people's tampering with emission control devices on late-model cars to counteract their adverse effect on performance, which negates emission control efforts by the manufacturer.[4] Such reports persist, despite a fine of up to $10,000 for removing or "rendering inoperative" any pollution control device. New cars in California use air pumps to reduce emissions further, and sample tests show a high failure rate, mainly because of owners' tinkering. Unless the cost of noncompliance is greater than the avoidable cost of compliance, there is little assurance that violations will be minimal.

The administrative costs of enforcing regulations that affect small economic units may be high. The probability of being caught and fined must be sufficiently large to motivate compliance. A recent report shows that the pollution control devices tend to malfunction, and many mechanics do not know how to fix them.[5] California and New Jersey, which have instituted statewide checkups, experienced failure rates of between 25 and 41 percent for 1970 and 1971 models. Neither owners nor mechanics may have incentives to maintain antipollution devices. A fine (or withdrawal of permits, as proposed by California) may provide the necessary motivation for mechanics to learn how to maintain these devices.[6]

If emission control devices easily malfunction or can be tampered with, frequent statewide automobile checkups may be necessary. There would be additional direct costs of regulation and large indirect costs, such as the loss of productive time during inspections and repairs. If emission control were a permanent or at least long-lasting operational characteristic of product design, inspection and maintenance costs for large numbers of small polluting units could be minimized. The balance between production costs and repetitive maintenance and inspection costs will be determined by available technology.

Monitoring and enforcement costs ought to be higher than they have been, in benefit-cost terms. What they would be, and what they ought to be, depends on the choice of pollution abatement policies. Presently, the principal policy is to require installation of pollution abatement equipment. The monitoring requirements to ascertain that this has been done are modest. However, installation is not enough; the equipment, whether catalytic converters or smokestack scrubbers, must function as designed if emissions are to be reduced as planned. Scrubbers require frequent and expensive maintenance in order to remove SO_x from smokestacks. The monitoring requirements for ascertaining that equipment is being used and functioning properly are much greater than those for determining that equipment is installed.

An emission tax, widely recommended by economists, would provide an incentive to the polluter to install and maintain pollution control equipment.

There would be no need for monitoring equipment whose installation and maintenance are optional. However, if an emission tax is to function as an adequate motivation, then emissions must be monitored. Such monitoring might be economically feasible only for major polluters. Ideally, monitoring would be continuous, which might require considerable expense not only for installation of monitoring equipment, but for proper maintenance of it.

Indoor pollution is not regulated (some industries excepted), although much of our pollution exposure occurs indoors. It is infeasible, either economically or politically, to monitor air quality inside many millions of homes and offices. Harmful emissions are controlled indirectly to some extent: housing codes require filters for heating and cooling systems and adequate ventilation for kitchens; regulations ban the production of some harmful sprays, paints, and other materials that contribute to indoor air pollution. There are no estimates of costs associated with enforcing these diverse indirect measures for reducing indoor pollution. There are limits to what can be done under our laws. The most harmful single indoor pollutant is tobacco smoke. Although prohibition of smoking in many public places does occur, prohibition at home seems out of the question.

Compliance Costs

The Council on Environmental Quality estimated total expenditures on air pollution abatement for 1971 at $2.1 billion, not all of which was related to health. Most of this cost, $1.7 billion, was for operation and maintenance. Expenditures in 1975 attributable to federal legislation were $9 billion, $5.3 billion of which was for operation and maintenance. Estimated costs for 1984 (in 1975 dollars) are $18.6 billion, with capital costs totaling $10.9 billion and operation and maintenance, $7.7 billion. (These are not total expenditures on air pollution abatement, but only that portion attributable to federal legislation.) In most years to date, expenditures for mobile sources (motor vehicles) account for over half of the total. The cumulative total for 1971-80 was estimated at $106.5 billion, and for 1975-84 at $136.5 billion.[7] The main difference between successive forecasts is a tendency for the capital-cost share to rise and for the mobile-sources share to fall.

EPA estimates of the costs of meeting state and national standards are much lower. The cumulative total for 1971 through 1979 was $47.3 billion, itself a large increase over a previous estimate (for 1970-78) of $23.4 billion (both in 1970 dollars). The most comparable estimate by the Council for Environmental Quality, for 1971-80, was $98.1 billion (in 1971 dollars). EPA estimates predicted annual costs in 1979 at $16.9 billion. Nearly half of the costs were for mobile sources.[8] None of these estimates includes the costs associated with plant shutdowns (capital losses, unemployment) and the complex costs involved in changing the location of economic activity.

The RECAT Committee's estimates of the annual costs of motor vehicle emission control are well above those of the EPA, although estimates are not strictly comparable. The EPA estimated costs for fiscal year 1978 at \$6.5 billion. The cost of extra required fuel exceeds investment and maintenance costs combined. RECAT estimated average annual costs for the conversion decade, 1976-85, at \$9.5 billion.[9] Both estimates assumed implementation on schedule of the 1975-76 motor vehicle emission standards. Irrespective of the severity factor used in weighting the relative effects of different pollutants on health, costs will exceed benefits. Estimates of costs (and also of benefits) should be reduced in line with postponement and relaxation of EPA emission standards for motor vehicles.

The NO_x standard is the one that imposes the greatest fuel penalty, as well as the highest investment costs, under current technology. The National Academy of Sciences estimated total costs of compliance for mobile sources under alternative assumptions as to the NO_x standard. The annual expenditures for automobile emission control would reach \$8 billion in 1985 with a 0.4-grams-per-mile standard for NO_x, but only \$4.7 billion with a 2-grams-per-mile standard. With technology in flux, there is some disagreement on the magnitude of the fuel penalty, which should decrease in time with new and improved techniques. Lee Iacocca estimated that the 2-grams-per-mile standard reduced fuel economy of 1977 Fords by seven-tenths of a mile per gallon.[10] Lead-free gasoline, required for use with catalytic converters and possibly required in the future by many states, seeking to comply with lead air quality standards (effective in 1982), also imposes a fuel penalty, variously estimated at 5 to 10 percent relative to 1970 emission standards.

The aggregate price impact of private investment in air pollution reduction has been projected by the EPA to result in a cumulative total increase in consumer prices of about 0.7 percent through 1977.[11] Over half of this increase would be caused by the 10 percent rise in the price of new automobiles as a result of emission control equipment and related costs. The impact of EPA industrial emission standards on unit costs and prices of most polluting industries is too small to bring about significant changes in relative prices or in the positions of the affected industries. This statement must be qualified with regard to electric power and nonferrous metallurgy plants, which have expected price increases of 4 to over 8 percent (see table 6-1).

More recently, the Council on Environmental Quality forecast a rate of increase in consumer prices of 4 percent by 1983, or between 0.3 and 0.4 percent a year between 1970 and 1983. It reported an increase of 1.6 percent realized by 1975, as well as an increase by 1975 of \$200 in the average price of new cars, attributable to clean air legislation.[12] (The consumer price increase

Table 6-1
Projected 1977 Price Increases in Stationary and Mobile Sources

Source	*Projected Price Increases (percent)*
Mobile sources	
New automobiles	10.0
New trucks	4.0
Solid waste disposal[a]	
Stationary fuel combustion	
Small and intermediate boilers	4.3
Steam-electric power[a]	
Industrial process studied	
Asphalt batching	3.0
Cement	2.2
Coal cleaning	1.5
Grain plants: Handling	1.3
Feed	0.1
Gray iron foundries	2.6
Iron and steel	1.2
Kraft (sulfate) pulp	0.7
Lime	1.8
Nitric acid	0.0
Petroleum products and storage	0.0
Petroleum refineries	0.0
Phosphate	0.5
Primary nonferrous metallurgy	
Aluminum	7.0
Copper	5.7
Lead	8.2
Zinc	7.1
Secondary nonferrous metallurgy	
Aluminum	1.2
Copper	0.1
Lead	0.1
Zinc	0.2
Sulfuric acid	2.4

Source: U.S. Senate, *The Economics of Clean Air,* March 1972, Table 5-1, p. 5-6.
[a]Not estimated because of lack of sufficient data

figures include the impact of legislation on water pollution and solid waste.) Recent estimates by General Electric (GE) of the cost of installing stack scrubbers in new electric utilities are 16 percent of the cost of new investment, expected to raise electric utility rates to households by 25 percent by 1990 (EPA estimates of investment cost and rate increases are less than one tenth of GE's).[13] The Council on Environmental Quality estimates air pollution control costs at 15 to 22 percent of investment costs of new coal-fired power plants, and operating costs at up to 7 percent.[14]

Early estimates of the costs of implementing the 1975-76 emission standards ranged from $350 to $500 per car (1971 dollars). The National Academy of Sciences (NAS) later estimated the cost at $600 (1974 dollars). These estimates raise the question of the impact of clean air legislation on the long-run use of cars. If prices had already risen $200 by 1975, with substantial further increases in prospect once the standards originally to become effective in 1975-76 are implemented, and with additional operating and maintenance costs, many people may reconsider their use of cars and the timing of their purchases, affecting the total number of cars bought and eventually the size of the motor vehicle pool. It is necessary in this case, therefore, to take into account among the costs and benefits the impact of emission controls on the automotive industry and on local mass transit and public transit in metropolitan areas. Price elasticities of demand need to be considered in evaluating the cost-effectiveness of air pollution control programs.[15]

None of these aggregate cost and price estimates includes the impact of new air quality standards for lead, to become effective in 1982, nor the shift to coal resulting from the great increase in the price of oil as well as from national policy to reduce American dependence on imported oil. The EPA has set no lead emission limits on any particular source, leaving it up to the states to attain air quality standards as they see fit. It is expected that lead and copper smelters will have to spend $600 million to clean up.[16] More likely, some of the smelters will avoid the control costs by discontinuing operations. The additional costs imposed by a shift toward coal (in fact, before the OPEC price increase in 1973 the trend had been away from coal) depend, first, on the rate at which consumption of coal is increased, and second, on whether the EPA will insist on use of expensive stack-scrubber technology, even for emitters burning low-sulfur coal and already meeting emission standards and those in low-pollution locations (where tall stacks could disperse pollutants, or temporary switches to low-sulfur fuel during periods of high air pollution could avoid failure to meet air quality standards). Over 85 percent of electric power generating capacity already uses low-sulfur fuel, so that EPA policy would have a large impact on costs.

Modest additional costs could be incurred to achieve large benefits not required by national legislation. No one proposes to denude the Smoky Mountains in order to reduce the emission of hydrocarbon aerosols, from whose presence they derive their name, nor to forbid roses in cities and suburbs because they cause allergies to some. Ragweed is another matter; it is an enemy to millions, loved by no one, and quite possibly eradicable from heavily populated areas.

Pollution Control and Economic Growth

Indirect costs of compliance include (apart from the effect of pollution control on prices), through the cost of production, a longer-run effect on growth because of diversion of investment and research from capacity- and productivity-increasing (and new-product development) activities to pollution-reducing activities. The Bureau of Economic Analysis of the Department of Commerce estimates business investment in air pollution abatement plants and equipment at $3.8 billion in 1977, or 3.4 percent of total business plant and equipment investment.[17] McGraw-Hill, whose estimate for 1977 is 13 percent higher, estimates that an investment of $292 billion is needed to meet all standards, (not just air standards) at all facilities. Industrial research and development for air pollution abatement totaled $690 million in 1976.

These cost estimates are incomplete. They are limited to the direct cost of air pollution control equipment and costs of operating and maintaining it (plus the associated research and development costs). They do not take into account the effect of such equipment on productivity and hence on unit costs of production. In the case of automobiles, there has been much disagreement on the effects of emission control equipment on mileage per gallon of gasoline; there has been less disagreement on the effects of unleaded gasoline. Pollution control equipment may also reduce productivity in many industrial processes.

This compliance cost remains to be assessed. These are clearly net costs by our current measurements of gross national product (GNP). Whether they would be net costs by some alternative measure that properly costs pollution and prices its reduction is not at all certain, however. Other trade-offs between pollution and growth include:

1. Costs of pollution abatement for the polluting industries:
2. Higher costs to all industries that use products of polluting industries, even those that are not themselves polluting;
3. Higher costs to all industries because of the higher cost of energy, varying with their energy intensity;
4. Emigration of highly polluting industries, reducing the effects of points 1 and 2.

Care must be taken in adding the costs of points 1, 2, and 3, since the second and third may consist largely of costs of abatement shifted forward. Most of the polluting industries, except for motor vehicle use, are producers of raw materials and intermediate products, or of inputs rather than outputs. Hence the effects noted in point 2 are pervasive. Insofar as sales of electric power are to households, then there is no input-related increase in production costs as they are currently measured.

It is often asserted that society is facing a choice between growth and environmental preservation. The conflict arises in large part from the failure to place a price on the quality of the environment. Thus, growth was achieved in the past at the cost of polluted air and water and defaced earth. But the polluters were not charged. A suitable charge on such activities, reflecting the social value of the environment, eliminates the conflict between environment and growth. Much of the environment that has been degraded—the air around, or most bodies of water—cannot be called private property. This is clearly evident for the factory smokestack whose emissions cannot be contained within the property of the factory, whatever its air rights may be.

In recent decades economic growth has apparently been associated with increasing exposure to air pollution, for growth was accompanied by the movement of many millions from rural areas to large cities, and by a rapid growth in population and labor force. Growth was largely quantitative. In the future, with the sharp decline in population growth and slowdown in the expansion of employment, growth will be more qualitative than quantitative: increased productivity, rather than just more output, will become important. The conflict between growth and environment will slowly ebb. The proportion of the population living in the most polluted areas seem likely to decline, as technological progress increases the efficiency of productive plants and equipment, leading to less waste, less input per unit of output, smaller residuals.[18] Much pollution, such as incomplete combustion, is an indication of waste. However, the contribution of changing patterns of growth toward moderating the conflict between pollution and production should not be overstated. It is growth in per capita income rather than in population that has accounted for most of the growth in emissions. The ongoing change in the composition of output, from goods whose production and/or use pollutes to services that do not, is often noted as an encouraging trend. It, too, should not be overstated. It is largely, if not entirely, a statistical figment. The productivity of goods-manufacturing industries has been rising and their prices have been falling relative to the service industries. As a result we employ an increasing share of the labor force in services and spend an increasing share of our income on them. But their output may actually have fallen relative to the output of goods. Since the output of service industries is very difficult to measure in many cases, we are not sure of past, much less of future, trends.

Assessment of the effect on economic growth is extremely complex and uncertain. Air pollution legislation and regulation has caused a roughly measurable increase in business investment to control pollution, and in operating and maintenance expenditures for the same purpose. To some extent investment in pollution equipment was and is at the expense of other investment; to some extent it may be in addition to other investment. Substitution maximizes the negative impact on growth, whereas supplementation minimizes it. Under conditions of full employment, investment in pollution must reduce other investment, or reduce consumption, or both. With idle labor and idle plants, it is possible for investment in pollution to increase the level of economic activity. In fact, the Council on Environmental Quality offers the view that pollution investment has not involved any growth penalty; on the contrary, it has increased growth during the recent years of recession and incomplete recovery.[19] This is an extreme view, reminiscent of Bastiat's famous Candlemakers Petition, which argued for measures against sunlight in order to increase employment. The legislation may have been followed by more employment and resource use, but investment that increases neither the output of consumer goods and services nor the productivity of labor and capital can hardly be a contribution to growth.

A second important question is the impact of pollution control investment on productive capacity and productivity. This investment is not intended to increase either capacity or productivity, which is why it may reduce growth. But it may actually reduce capacity and productivity, and reduce the effectiveness of all new investment. At one time, auto emission control involved a fuel penalty, and some fear that it still may. There is no question that lead-free gasoline does involve a fuel penalty. Stack scrubbers can involve substantial "down" time, effectively cutting capacity and productivity. There is no assessment of the total effect of current pollution abatement technology on the efficiency of the associated factors of production. It is this effect that makes pollution control investment incur a growth penalty, even if it does not displace any other investment and even if it employs resources that might otherwise be idle. Of course, in the long run there is no reason to assume that the level of economic activity would be higher with the EPA than without it. On the contrary, since the level of utilization of resources in a dynamic economy depends on its growth rate, pollution control expenditures are likely to reduce both the level of activity and the rate of growth in the long run. What the EPA does is to increase the output of clean air, but this is not included in the GNP, whereas the cost of cleaning the air is included among the costs of producing the GNP.

Since much of the technology is new or experimental, experience on which to estimate impacts is either lacking or inadequate. It may be expected that performance will improve in the future, in terms of efficiency both in reducing adverse impacts on output and productivity.

A further consideration is the long-run tendency for emissions to increase for a given level of technology, as a function of increasing output. At the most general level, energy output has been increasing approximately 5 percent a year. Even with a reduced rate of growth in the future, it is clear that emission technology and standards which are adequate today for achieving high national ambient air quality will prove inadequate soon enough. Although one may expect some change in the structure of production away from emission-intensive products and energy-intensive goods and services, as well as slower population growth, output of electric power and of most polluting industries will continue to increase. Thus, we can anticipate ever more stringent emission reduction requirements, which will time and again threaten to slow down growth.

The assumption so far has been that producers will comply with air quality and emission standards by investing in antipollution equipment and by changing inputs or production processes. Another response is to discontinue production, if costs of compliance are too high. The economic life of plants and equipment is curtailed, jobs may be lost and skills wasted, and communities may be adversely affected. If only new plants and equipment are required to comply with emission standards, this response is minimized. A variant is to stop production in one location and start or expand production in another, possibly foreign, that has less demanding standards. No one knows the magnitude of any of these effects. Emigration of industry depends on foreign production and distribution costs as well as on environmental regulation. Its short-run consequence is to reduce capacity, output, and growth, although in the long run it might merely accelerate the international redistribution of industries that is going on anyway.

Before the oil price increases and the embargo in 1973, it was possible to regard with some equanimity the consequences of the EPA for economic growth. There would be difficulties initially, as large investments were made in a short period of time to comply with EPA regulations—investments that neither increased productivity nor expanded output. Once all polluting sources were suitably equipped, then there would be only maintenance and replacement to worry about. But the energy crisis and the responses it has and will induce radically alter the picture. The major response for the foreseeable future is a greater reliance on coal; and the combustion of coal is far more polluting on the average than the combustion of oil. The greatest impact will be in the electric-power generating industry, which relies on coal for over half its output. The increase in coal production proposed in President Carter's energy program is sixty million tons a year. Any program eventually implemented must for a good many years place stress on greater utilization of coal. Among the longer-term prospects are gas and oil derived from coal by processes likely to generate considerable air pollution, greater than that associated with production of petroleum.

To attain the same air quality in the future will require far more investment than was necessary before the national policy to reduce our dependence on imported oil. How much more depends on what actually happens to plans for conversion to coal. Of course, since the costs of given levels of air quality have increased as a result (whereas the benefits have not changed), it would be reasonable to arrive at a compromise, a somewhat higher level of ambient air pollution, as the optimum. Coal, pollutant emissions aside, is a cheaper source of energy than oil. The large diversion of research and development to providing new sources of energy and to finding ways of reducing the pollution associated with coal burning and coal mining is perhaps more harmful to growth than the large diversion of investment toward conversion and pollution abatement.

Foreign Trade and Payments

It has been claimed that clean air legislation is a handicap for United States exports and an advantage for foreign exporters to the United States. The possible effects are limited to those few products whose prices may be significantly affected by legislation in the United States and which are internationally traded. So far as the polluting characteristics of final products are concerned, imports have to meet the same standards as domestic products (e. g., emission standards for automobiles). Thus, only standards applying to the manufacture of final products or to intermediate products could possibly affect United States foreign trade, or provide a stimulus for investment and production abroad instead of at home.

The possible effects depend also on what foreign competitors may do about their own emissions. It is likely that the less developed nations will remain less concerned about air pollution problems than about economic growth and export expansion; thus, export of some American goods would suffer, imports of some goods from less developed nations would increase, and some investment might be diverted from the United States to developing nations.

It is an entirely different matter with advanced industrial nations. Although the United States has been ahead of many advanced nations in enforcing ambient air standards and emission controls, other advanced nations can be assumed to adopt similar standards soon if they have not already; health may seem more important to them than maintaining past growth rates. Most advanced nations face a worse pollution problem than the United States because they have much smaller land areas, much higher population density, and a greater concentration of polluting industries. This is particularly true of Japan whose GNP per acre is approximately seven times that of the United States. What this means is that stricter emission controls are needed in Japan and some European nations than in the United States to attain the same air quality. The greater concentration of population also means a higher rate of population exposure to pollutant emission than prevails in the United States. Thus, on a net basis the United States is not likely to be disadvantaged by and could conceivably gain from a rapid international spread of air quality standards and emission restrictions.[20] On a net basis the terms of trade (the ability of United States exports to buy imports) might rise slightly as a result of an improved competitive position in international trade.

There is a tendency to blame environmental legislation for the migration of some of the more polluting industries, which is taking place largely as a result of cost differentials unrelated to environmental law and of the industrialization of raw materials-exporting nations. There is no convincing evidence that expected costs of compliance with clean air have been decisive, although they may have accelerated the migration somewhat. The effect on the United States would depend on the level of exports of the migrating industries (which would be lost), their import content (which would be eliminated), and the level of U.S. imports of their products once they have relocated, a function of delivered prices (including transportation and tariffs). Clearly the effect would be adverse, but it is hardly likely to be significant. On the other hand, the shift toward coal, raising domestic costs of pollution abatement, might have a net positive effect on trade by reducing oil imports. One is tempted to say that electric utilities, unlike smelters, cannot relocate to countries that are less concerned about emissions. This may not be entirely correct. Californian utilities may consider construction of new plants in Mexico to circumvent costly delays and possibly insuperable obstacles to their construction in California or nearby states.[21]

Technology and Time-Cost Trade-offs

The goals toward which the nation should aim are determined by the relation between ambient air pollution and health. The standards for pollution emitters are determined by the complex relation between emissions and ambient air pollution. *When* the goals and standards should be attained—the rate at which they should be approximated—is an entirely separate issue. The speed with which air pollution is reduced will affect the cost of pollution abatement, both in total and at the margin.

One estimate of the annual cost of stricter air quality controls gives a range of \$12 to \$18 billion over the 1977-87 decade.[22] Three fourths of the range, or \$4.5 billion annually, is accounted for by the premature obsolescence of capital stock as a result of enforcing stricter environmental controls as scheduled, without deferral. The remainder of the range depends on the rate of technological progress in abatement and in energy production and the rate of cost increase in energy resources. The excess costs of implementing standards before 1985 have a present value ranging from \$56 to \$252 billion, depending mainly on the date of implementation, but also on the above-mentioned factors and the discount rate. The National Academy of Sciences estimated that a delay of the 0.4 ppm NO_x standard from 1978 to 1980 would save \$6.9 billion by 1985.

The conflict is between accelerating the achievement of air quality goals and the higher costs of acceleration, assuming there are no benefits in reducing air pollution more than is called for by the goals. If, however, further reductions in air pollution have value, then a third consideration enters the calculation: How much emission reduction is possible (and at what cost) with alternative technologies that will become available at different points in the future?

New equipment is less polluting than old. If the average life of equipment is five years, improvement can be introduced rapidly by the normal process of replacement. If the average life of equipment is twenty to thirty years, even if new equipment has the desired low-polluting quality, measures in addition to normal replacement may be needed to accelerate improvement in air quality.

The average automobile has a life expectancy of ten years. For the industrial or commercial furnace, life expectancy will exceed twenty years. Thermoelectric power plant equipment has even greater longevity. The lag between best and average performance, therefore, will be least for motor vehicles and greatest for electric power plants, so there is less call for emission control standards on old cars than on old furnaces. In fact, EPA motor vehicle standards refer only to new cars, though states with acute problems, for example California, consider standards for old cars as well.

The assumption above is that pollution is best reduced through improvements in the design of the combustion chamber and associated equipment. In some cases, auxiliary equipment may be the most cost-effective approach, and may be as easily added to old as to new equipment. Here the replacement rate is irrelevant. If the major way of reducing pollution is through changes in fuel, then progress in pollution reduction depends on the need for new equipment to produce low-pollution fuel and on the replacement rate of this equipment (or of the former equipment). Accelerated replacement of long-lived equipment is a calculable cost of accelerated pollution reduction.

When knowledge about air pollution and its effects on health is inadequate, the policy to be recommended must have a great deal of flexibility. Air quality standards must be subject to reconsideration. Time schedules for their attainment and specific regulations to implement such schedules must also be reviewed as new information warrants. On the other hand, standards, deadlines, and regulations which may be changed in the near future are very costly to implement. Billions of dollars of investment are involved, and the resulting physical capital may have no alternative uses. Thus, a change of policy may render a great deal of capital equipment useless, or at least unnecessary. Flexibility and responsiveness to new and better information can come at a high cost. The great increase in the price of oil may affect policy by changing the trade-off between emissions and energy.

Policies are implemented in durable capital. Once the investment in a particular technology has been made, it may take considerable time to alter. In terms of optimizing antipollution policies over a longer period of time, trade-offs between pollution in the near-term future and pollution in the more distant future must be calculated (or will be implicit in decisions). Premature freezing of technology in durable equipment while the technological learning curve is rising may mean greater pollution over the longer period, because less effective techniques were embodied in capital equipment than would have been embodied had choice and investment been delayed. Technological forecasting becomes an important ingredient in determination of policy and the timing of its implementation. It is always true that a commitment this year may turn out to be unwise because of an unexpected breakthrough next year. The reference is not to this type of occurrence, but to technological progress that can be anticipated.

In its semiannual report on January 1, 1972, the Committee on Motor Vehicle Emissions of the National Academy of Sciences–National Research Council found that the "technology necessary to meet the requirements of the Clean Air Amendments for 1975 model-year light duty motor vehicles is not available at this time." They state further: "If enforcement of the requirements of the Act for 1975 vehicles were deferred for one year, the opportunity thus provided for further development and field testing would enable manufacturers to significantly improve the performance and reliability of vehicles equipped to meet the requirements in the hands of the customer."[23]

Although the 1973 report of the Committee on Motor Vehicle Emissions of the National Academy concluded that four types of emission control systems would meet the prescribed emissions standards for 1975 "during certification testing," and that vehicles incorporating these systems could be mass produced to meet expected demand for that year, the committee was unable to conclude that such systems would continue to meet emission standards for a reasonable period of time.[24] It had no data concerning deterioration of emission control systems under normal wear-and-tear conditions of customer use.

The time constraint imposed on auto manufacturers by the 1975-76 deadline forced them to concentrate their development efforts on the technologies most likely to be ready by the deadline, even though they might be inefficient and costly. This diversion furthermore may deter development and introduction of superior alternative techniques for a long time to come. In fact, the auto industry has concentrated on minor modifications of the conventional internal combustion engine and the addition of catalytic converters. Other technological alternatives have not been pressed because they require a long development effort before mass production is feasible.[25] The postponement of the 1975-76 standard a year or two at a time, with limited advance notice, is not conducive to development of superior alternative technologies with longer and less predictable lead times than the catalytic converter.[26] It may even encourage a belief that the practice will continue indefinitely, and that the development of the needed technology can be deferred or pursued in a leisurely manner.

Quite apart from the influence of time pressure on the choice of technology is its effect on the cost of developing a given technology. Crash programs are not compatible with cost minimization, nor with optimal initial product performance and reliability. One by-product of the pressure to develop efficient catalytic converters was the skyrocketing price of platinum, the demand for which suddenly expanded greatly. It is impossible to predict how long a research and development effort ought to take, or how much it ought to cost, so that judgment is needed to avoid excessive time pressure on producers, on the one hand, and too much permissiveness, on the other. Policies other than emission standards—for instance, emission taxes (to be discussed later)—might be more appropriate for optimizing the rate of technological progress in pollution abatement.

The use of catalysts, the strategy adopted by the automobile industry to meet the 1975-76 emission standards, raises several difficulties. It requires unleaded gasoline, since ethyl lead destroys platinum and palladium catalysts. Unleaded gasoline, apart from the very large investments required to produce it in requisite quantities, is less powerful than leaded gas. This means additional fuel consumption, which to a very limited extent counteracts the reduction in pollution achieved by the catalyst. More important, added fuel demand is very untimely when energy shortages, and particularly oil and gasoline shortages, are threatening.

Estimates of the additional fuel consumption required by conventional engines to meet present emission standards range up to 30 percent. The Wankel engine, which may be able to meet 1976 standards for most pollutants without modification, would use even more fuel. The stratified charge engine, on the other hand, uses little fuel and generates little pollution. It is not, however, ready for mass production. Diesels have long been operational and might meet standards for most pollutants, but not for NO_X, and they also produce harmful aldehydes. Electric cars do not pollute, but will have very limited range and burdensome recharging needs until an adequate service network is available for them. They consume coal indirectly and are very energy-wasteful, because of the cumulative energy lost in converting heat to electricity and its transmission and final conversion to mechanical power.

Also, to meet federal 1975-76 standards for emission, the auto manufacturers asked the oil companies to produce gasoline with low enough volatility to keep pollutants below the required level.[27] Union Oil Company estimated that meeting the desired volatility restrictions would require an additional 250 million barrels of crude oil per year. The production on a crash basis of enough unleaded gasoline to meet 1976 requirements is estimated to increase costs 1.8 to 4.7 cents per gallon, or up to $4.7 billion annually. On a phased-out basis, the cost increase would be an estimated 1.0 cent per gallon.[28]

The catalysts themselves pollute. Catalysts being tried out incorporate chromium, nickel, rhodium, ruthenium, and other highly toxic elements, in addition to platinum. Such catalyst compositions have been alleged to emit fine particles under conditions simulating operation of automobile engines.[29] Furthermore, it was feared that platinum-based catalysts promote the conversion of gasoline sulfur into sulfuric acid mist in unacceptable concentrations.[30] Recent tests, however, indicate that these fears are unfounded.[31] In addition, gasoline manufacturers will increase the proportion of olefins and aromatic hydrocarbons, which are more harmful than other hydrocarbons in gasoline, to counteract the reduced octane rating of unleaded gasoline.

Another example of the adoption of a probably inappropriate technology through haste to reduce pollution is the use of stack scrubbers to remove SO_X from flue gas. The process generates one ton of sludge for every three tons of coal burned, creating a disposal problem which can rapidly become unmanageable, particularly for electric utilities in urban areas. It is also an expensive and currently unreliable process. But the insistence on immediate adoption of the best available technology will saddle coal-burning power plants with capital costs and disposal problems for years, and reduce their productivity because of the need for shutdowns for cleaning and maintenance. Furthermore, it will induce them to buy cheap high-sulfur coal. Meanwhile, development of superior techniques, including the cleaning of coal before burning, is retarded and discouraged.[32]

The question to be considered, even if it cannot be answered with precision, is how much waste is involved in adopting short-term technology that has long-term effects. How much waste is associated with the rush to meet an imminent deadline? Are the benefits to be derived from meeting a desirable emission or air quality level in haste worth the extra cost?

A final issue involving trade-offs over time refers to the nature of the pollutants involved. Some have lasting and cumulative effects. Others may have temporary, reversible, noncumulative effects. It may be wise to postpone reductions in the latter type of pollution in order to make inroads on the former. Pressure to hasten technological progress involves additional costs. Whether or not these costs are worth the benefits depends not only on how large the additional costs (of a deadline demand) are, but on whether the benefits achieved thereby are cumulative over time or not. In this regard, too much stress has perhaps been placed on reduction of CO emissions. The higher temperatures and improved combustion that reduce emissions of CO and relatively harmless hydrocarbons increase the production of more dangerous nitrogen oxides.

Evidence that the measured effects of some air pollutants on health are reversible indicates that it is better to postpone antipollution measures until better techniques are available, rather than to sacrifice future performance by rushing into the best techniques currently available. This conclusion must be regarded as quite tentative. There is evidence, certainly, of long-term effects of cumulative exposure and of some nonreversible health damage (apart from mortality). To calculate the optimal strategy we would have to know how much better future techniques will prove to be than techniques currently available, and we would also have to know how much longer we would have to wait for the better technique. This precision in technological forecasting escapes us now, and probably always will.

The cost and feasibility of emission control, if not the level of emission itself, can be affected by the choice between existing sources of heat and power. Electric heating and electric propulsion involve a shift from methods of fuel combustion that are difficult to monitor and expensive to control with regard to air pollution to methods which are much easier to monitor and technically much easier and cheaper to control. The basic reason is economies of scale and reduction in the number of relevant units. Regulation and monitoring are required only at the stage of generation of electricity. But electricity is energy-wasteful compared to direct burning of fuel to produce heat or mechanical energy.

Costs and Benefits Compared

In attempting to compare costs and benefits of air pollution abatement, we are faced with several complications. One is that the estimates of benefits are very rough. Another is that there are benefits other than health, and these may be even more difficult to measure. In general, only direct monetary benefits associated with abatement of air pollution to human health, materials, and vegetation have been estimated. Even so, these studies rely heavily on value judgments, are incomplete, and cannot be assumed to come anywhere close to the actual damage costs. There are many intangible or nonpecuniary costs that cannot be measured. Beyond the fact that they are not nonexistant, very little else is known, even about their order of magnitude.

A third complication is that abatement cost estimates are quite incomplete, being limited essentially to costs of investment in and maintenance and operation of pollution abatement equipment, ignoring impacts on productivity and economic growth. Fourth, costs and benefits predicated on implementation of standards as originally scheduled have been rendered obsolete by repeated postponements, possible reconsideration of some standards, and promulgation of new rules, and will be further affected by the new air quality standard for lead. The additional time allowed for technological development in motor vehicle emission control will reduce to an unpredictable extent the future costs of abatement and may permit reconsideration of standards.

Furthermore, there are joint costs as well as joint benefits of improved health, reduced damage to property and vegetation, and other possible correlates of air pollution. Also, the emergence of the energy crisis will increase both future costs and future benefits of air pollution abatement. There is a sharp conflict between the goals of energy self-sufficiency for the United States and air pollution abatement, reducing the priority of clean air on the agenda of the nation.

The most serious problem is how to measure benefits and costs in terms that will allow comparison not only in total, but at the margin: comparison of increments (decrements) in abatement costs with the associated changes in benefits. This is necessary in order to arrive at optimum policies and optimum standards. The various types of pollution damage costs cannot be added since they are interrelated. Declines in land value may reflect subjective impacts on health and environment and/or the costs of counteracting these impacts. Effects on land value, therefore, cannot be added to any other effects without some double counting. Damage and avoidance costs are substitutive, not additive. It is questionable that effects on health, aesthetics, or quality of life can be adequately or reasonably expressed in a common denominator that allows their addition to or comparison with abatement costs. There are limits to the economic calculus as a standard for comparability, a measure of all values.

In estimating whether pollution control measures are worth the cost, and the level of pollution control at which additional control costs are just equal to the cost of the damage they avert, it is not correct to sum all costs associated with reducing pollution. These costs include: (a) costs which are incurred only because of federal legislation; (b) costs which would have been required by state and local governments even in the absence of federal legislation, some of which are not required by the federal government (Pittsburgh and St. Louis, for example, cleaned their air long before 1970; the state of California imposes stricter controls on motor vehicle emissions than does the federal government); and finally (c) some costs would have been incurred in the absence of any legislation at any level of government. Immobile plants may in some cases be sensitive to the quality of environment in which they function and their employees, including management, must live. Some pollution-reducing expenditures are undertaken because they are profitable. The tremendous reduction in the amount of coal required to produce a ton of steel greatly reduced emissions in the steel industry, although that was not its purpose. Improved combustion cuts down on emissions and on fuel consumption.

Those expenditures whose impact on emissions is secondary to other objectives, or which would have been undertaken voluntarily by the polluter or others exposed to its emissions, should not enter the benefit-cost calculation. In fact, if we employ marginal costs and marginal benefits to determine optimal pollution abatement, they do not. If we use total costs and total benefits, they do.

It is not always easy to distinguish between expenditures whose main purpose is to reduce emissions and expenditures whose emission-reducing effect is a secondary objective or by-product. With tax and regulatory incentives to invest in pollution-control technology and with favorable publicity to be gained from doing so, it is inevitable that such expenditures will be somewhat overstated.

Given the great uncertainties surrounding estimates of the benefits of air pollution abatement, and substantial uncertainties surrounding the costs (as suggested by the large changes that have taken place in EPA estimates), no numerical estimates are possible on the relation between costs and benefits. About all that can be said is that benefits appear to exceed costs by a comfortable margin for industrial (including electric utility) pollution abatement, and that the benefit-cost ratio is much less favorable for mobile sources. Given our present knowledge of the damage done and of the efficiency and cost of available pollution abatement techniques, it is clear that emission controls on sulfur compounds and particulate emissions from electric utilities and industrial polluters are justified. How much of a constraint is justified is another matter. Our knowledge is not sufficiently precise to judge whether constraints are excessive, inadequate, or just about right. On the other hand, the evidence is not clear that in most places the benefits from reduction of CO and hydrocarbon emissions, auto-related pollutants, are worth the costs; the tentative inference is that incremental benefits are less than incremental costs. The ultimate emission standards for NO_x which account for a large part of the eventual cost of auto emission control, clearly are not worth the costs at this time.

A plausible case can be made that reductions in emissions from mobile sources of the order envisaged by the Clean Air Amendments of 1970 are likely to cost more than the estimated value of the resulting benefits. They would clearly have exceeded benefits had the standards been implemented on schedule. The repeated deferment of standards, although not conducted in such a way as to allow the best use to be made of the additional time (more in the spirit of a series of installments in the Perils of Pauline), nevertheless have greatly reduced the costs incurred so far and those to be incurred in the future. Although benefits have been reduced, the benefit-cost ratio has been improved. Whether the marginal benefit exceeds the marginal cost with these deferments, or even whether the total benefit exceeds the total costs, is not proven.

This unsatisfactory verdict is more the result of the way in which air quality standards are implemented than of the standards themselves. If emission standards were not invariant across the nation but were adjusted to reflect regional differences in air quality, the costs of obtaining a given benefit could be greatly reduced, not only for mobile sources, but for stationary sources as well.

Energy policy stressing coal will raise the cost of attaining any given level of ambient air quality, especially for SO_x and particulates; but it will also raise the benefit, since it would result in worsening air quality in the absence of abatement efforts. A reasoned compromise between clean air and energy policies has not been attained, and cannot even be attempted until there is a clear and effective energy policy. The conflict between clean air and economic growth is not as simple or evident, but it exists and must eventually also lead to at least implicit compromise.

Whatever the present calculus of costs and benefits, we can rest assured that in time new techniques will improve the effectiveness and reduce the cost of pollution abatement to the point that standards now seemingly unjustified will prove worthwhile. The faith of Congress in the technological prowess of America was not misplaced; Congress was just too impatient in expecting timely solutions.

"Clean air legislation constitutes an almost classic example of policy outrunning knowledge and research. To determine the costs and benefits of almost any program requires a synthesis of a large amount of different kinds of information, often physical as well as economic in character. . . . This type of information is only beginning to be developed."

7 Geographical Distribution of the Costs of Clean Air

Impacts of Alternative Policies

Differences in level and composition of air pollution within the United States are very large. Differences by type of pollutant source are important. The local costs of clean air depend upon the target: national air quality standards, regional nondegradation policies, offset rules. If the goal is a national ambient air standard, then in most areas the cost is zero, although it varies widely, being very large in New York and in some industrial centers and small in most cities. Polluters in the worst areas will see their costs rise and their freedom to function restricted. This is true whether they are affected by stricter emission standards directly or only indirectly, through higher taxes and the adverse impact of higher standards on other local business, including temporary reduction of polluting activities during hazardous air alerts. The result of changes in relative costs of different areas will be some inter-area redistribution of employment, income, and population. The East and California will suffer relative to the South and Midwest. Large metropolitan areas will suffer relative to smaller cities.

If, on the other hand, no environmental degradation at all is to be permitted, the cost is positive everywhere, and could be highest in areas with clean air. The reason is that dirty cities can fairly readily reduce air pollution by restricting emissions of the most flagrant offenders, often quite cheaply. Clean air areas do not have any way of easily and cheaply improving the quality of the air to counter deterioration caused by new sources of pollution. New industries or additional population must therefore be Simon-pure nonpolluters. The already dirty areas, which can trade off reductions in existing industry against additional contributions to pollution by new industries or firms, have an easier time. This disadvantage for areas with clean air may not apply relative to dirty areas that do not meet national or state standards, and that are therefore required to improve air quality.

It is not necessarily true that all areas with clean air lack the option of reducing emissions from existing industry or population as a trade-off against the pollution costs of economic growth. Nevertheless, it is clear that the nondegradation policy is a zero-growth policy for most clean air areas, which incidentally accelerates the growth of highly polluted areas.

Clean air laws and EPA rules have threatened to stop growth in many areas and to slow growth appreciably for the nation. In 1975, 156 of 247 air quality districts failed to meet the goal for at least one pollutant.[1] Although by 1977 all but twenty-nine met federal ambient air quality standards, these twenty-nine included fifty million people and a large share of the industrial activity of the nation.[2] They constituted 90 percent of the nation's monitored air quality regions. A strict adherence to the law would have meant that these areas would not have been able to add any polluting plants. Their growth, even their status quo, would have been maintained only with great difficulty, for new plants are needed as old ones close down or move away. Nor was the option of shifting all new plant investment to clean air areas open. Class I areas, those with least pollution, were not allowed significant deterioration in air quality, which would have been the consequence of significant economic growth. Thus, economic growth would have been limited to areas which fell in neither category, and would inevitably have been reduced as a result.

To escape part of the dilemma of its own devising, the EPA instituted a policy establishing "off-set rules" for licensing new sources of pollutant emission in areas that did not meet national ambient air standards. In order to obtain permission to add more than one hundred tons of pollutant a year, a plant must arrange for an off-setting reduction of at least as much, and in some cases has been required to obtain a reduction twice as large as its own emission.[3] This rule is expected to lead to some unintended and undesirable results. If a firm does not have other plants in the area which can provide off-setting reductions in pollutants, it must buy an off-set. Thus an incentive is created to avert emission reduction and to hoard off-sets, which now have a substantial market value, until a buyer is available. Local and state governments may also face perverse incentives to procrastinate on enforcement in the interests of economic growth, until a new polluter or a polluting plant expansion is available.

The necessity of obtaining off-sets raises the cost of new locations or significant expansions in dirty areas. It will raise the costs even more in areas where off-sets are hard to get and therefore have a scarcity value. A nondegradation rule is equivalent to a universal off-set policy, which raises the costs of location and expansion in inverse relation to the level of pollution in an area. However, off-set rules do not raise the costs of existing polluters (not do nondegradation rules). On the contrary, they enable existing polluters to sell "pollution rights" to new entries and expanding firms. Since the condition of selling pollution rights is reduction of pollution by an amount equivalent (or more than equivalent) to the rights sold, existing firms for whom pollution reduction is cheapest will be the ones most likely to sell their pollution rights. They will be able to shift most if not all their pollution abatement costs to new firms, and in some cases will make a profit as well. Shifting the costs of pollution abatement for old plants to new investments that are already subject to emission standards might well discourage new investment and prolong the economic life of old plants and equipment. Since new plants and equipment tend to be less polluting than old, such a consequence would run counter to the objectives of clean air legislation.

The EPA has gone further, allowing exceptions to its off-set rules for areas making "substantial progress" in reducing air pollution to meet national standards. Since at this time neither the interpretation to be placed on "substantial progress" nor the magnitude of the exceptions to be allowed is clear, it is not possible to judge the consequences either in delayed attainment of national standards or in distribution of economic growth between dirty areas and other areas.

The consequences for interregional distribution of economic activity depend on exceptions to air quality standards for areas not meeting them, on the way in which off-set rules are used, and on the future of nondegradation rules for clean air areas. But clearly there will be some dispersion of polluting industry from areas of concentration and large cities. Since choice of new locations will be constrained by air quality rather than exclusively by profitability criteria, one may expect some loss of efficiency. Furthermore, the more dispersed locational pattern is likely to increase needs to use energy for transportation. It thus runs contrary to the requirements of energy conservation and self-sufficiency.

National emissions standards, rather than ambient air standards, would affect all areas in proportion to their concentration of pollution emitters, without affecting their relative position. New coal-fired power plants are a case in point. They must install the latest available technology—in this case, stack scrubbers—to reduce air pollution, wherever they may be located. This will also be true of utilities converted to coal from other fuels in furtherance of national energy policies.

Since most polluting industry is subject to national emission standards as well, it may find no difference in emission control costs in alternative locations, as long as it does not have to buy an offset in addition to controlling its own emissions. Thus, even if nondegradation rules are relaxed substantially, there would be no advantage to the business that is subject to emission standards in locating in cleaner rather than dirtier areas. The northeastern and midwestern industrial belts will clearly be the regions most adversely affected by enforcing national ambient air standards. They have the greatest concentration of most-polluting industries and depend almost exclusively on coal-burning electric power. California, more for climatic reasons than for its concentration of polluting activities, will also be severely affected. These areas have the most to gain from postponed enforcement of air quality standards, widespread use of off-sets, and strict nondegradation policies for clean air areas. Conversely, the other regions, especially the West (excepting San Francisco and southern California), are adversely affected by strict nondegradation policies and by uniform national emission standards, and would gain by strict enforcement of national air quality standards.

Regulatory Uncertainty

The uncertainties associated with existing administrative discretion are pervasive. What areas that do not meet the standards will be permitted to have new polluting firms, or will be spared draconian measures to reduce pollution such as the proposed 50 percent reduction in motor vehicle traffic in New York City? What, indeed, constitutes failure to meet standards? Cities may exceed all the annual standards or just one, or meet all annual standards but exceed shorter-term standards for one or more pollutants, one or many times. For what cities will off-sets be required? Will a new firm generating a pollutant for which the city meets the standard be denied entry because the city is in violation of the standard for another pollutant? Will an off-set be variously stipulated as the equivalent amount or, as has already happened, twice the equivalent amount, or some other multiple? Must the off-set be in the same pollutant(s) generated by the new or expanded firm? If not, how is equivalence to be established? What index of relative severity of damage will be used to determine what reduction in particulates is equivalent to an increase in SO_2? How is satisfactory progress toward compliance with the standards to be defined, now that nearly every deadline has been changed at least one time? What areas will be allowed to degrade air quality and by how much?

Despite the numerous retreats of the EPA in enforcing the original legislation, expansion of dirty areas is undoubtedly discouraged to some extent because of the revealed lack of certainty. It is in these areas that regulatory and administrative uncertainties are greatest.

What Ought To Be Done

The cost of reducing air pollution is a function of emission sources as well as of ambient levels of air pollution. The two are related, but not closely. The most cost-effective way of reducing pollution by a given amount is to concentrate on the dirtiest polluters irrespective of their location. But since the ultimate objective is to reduce effects on health, rather than pollution abatement, the most cost-effective way to achieve such reduction by a given amount must encompass both the increasing costs of emission reduction and the decreasing benefits for health, as well as initial area levels of pollution (and health effects). The levels of costs and of benefits will vary from area to area, as will the ratio between the two. Therefore, the optimal extent of pollution abatement will also vary. Currently, although per capita expenditures on pollution abatement vary widely between areas, they are based entirely on concentration of polluters required to reduce emissions, without regard to air quality in the area.

Typical state implementation plans, which consider neither pollution emission control costs or enforcement costs by area, prove to be far more costly than alternative strategies for attaining the same air quality. The least-cost strategy considers both the marginal emission control costs and the emission dispersion characteristics of an area. A strategy aiming at a given emission reduction at least cost considers the marginal emission control costs of each individual pollutant, but assumes a constant relation between emission reduction (regardless of source) and ambient air quality. Although more costly than the strategy that also considers dispersion characteristics of emissions, it is far less costly than typical state implementation plans.[4] Indeed, considering the benefits associated with pollution (growth in income, employment) as well as the costs, some areas would not curtail emissions at all, but allow the number of emitters to rise and air quality to fall.

Considering all costs and benefits, the optimal air quality standard will vary substantially between areas. This would be the outcome of equating the marginal costs of abatement with the marginal benefits. Ideally, each area would be free to choose the most cost-effective way of achieving this standard (in fact, the standard would depend somewhat on the cost of the selected strategy). Each would have some discretion to vary emission standards and coverage wherever it can be done without influencing abatement costs substantially. In particular, this approach argues for a two-car strategy, with less stringent emission standards for cars sold in low-pollution areas. Given actual cost and benefit functions, economic and equity criteria point in the same direction. The dirtiest areas should cut down emissions by a larger absolute amount and for a larger proportion of the universe of polluters. Areas with low levels of pollution should ignore many polluters who would be constrained elsewhere, and curtail emissions less among those polluters that it does regulate. These conclusions, it should be stressed, are concerned only with health benefits. Areas have other concerns and differ widely in the services they offer their citizens and the nation. Levels of pollution, or even increments in pollution, that might be acceptable in cities that are steelmakers and butchers for the world may be intolerable in Yellowstone Park.

We have seen that the main source of all pollutants for which ambient air standards have been set are natural, not man-made. Thus, emission control is only a sensible policy in those places where man-made emissions are very important relative to total air pollution. This may not be true of many Class I areas. Current legislation does not affect emissions of natural origin. Indeed, only natural emissions associated with soil management and vegetation have been influenced by man, who has sought to control wind and water erosion by ground cover, protective tree belts, and contour plowing, and who has affected vegetal emissions by an influence on the amount and type of vegetation and on the incidence and control of grass and forest fires.

Emission standards might be varied by area; areas with relatively pure air or low total man-made emissions would be excluded from control, and those with dirtier air would be subject to stricter emission rules. As the number of emission sources increases, stricter and stricter standards would be applied, and more areas would be regulated. Difficulties arise when emission characteristics are built into equipment at the national level. It is too costly to build equipment with multiple emission standard characteristics, and there are administrative problems in containing mobile sources (cars) within the metropolitan area to which they are shipped and in which they are sold, if there are substantial cost differences between areas. But two standards are manageable and greatly reduce the difficulties of implementation. In fact, stricter emission control devices are built into cars for sale in California than into cars sold elsewhere. In the case of car exhaust, emission standards are not completely built in. Localities can still adjust to their particular circumstances by varying frequency and standards of inspection.

In the case of mobile sources, there is no possibility of separation of people and sources of pollution. In the case of electric power plants that burn coal, some spatial separation is possible through location in coal fields. Other sources of emission that are not ubiquitous, as is the motor vehicle, or highly dispersed geographically, lend themselves to a policy of spatial segregation. Whether this can be an important or a good strategy is another matter. But it is feasible to locate new smelters, cement plants, and pulp and paper mills away from large concentrations of population, thus reducing pollution exposure without reducing emissions.

Much pollution control is not built in but added on to productive equipment. In either case, there may be alternative technologies, with varying effectiveness and varying costs. However, polluters have been given little choice. First, they must meet national emission standards even if they are located in areas with far better air quality than national standards demand. Second, deadlines have not allowed for technological choices involving substantial development time. Thus, automakers were constrained to use catalytic converter technology. New coal-burning utilities are required to use the best current technology, which means stack scrubbers. Clean air areas, therefore, have not obtained a cost advantage for new investment associated with differences in air quality; incentives for redistribution of polluting activities have been minimized.

Nondegradation rules are clearly in conflict with the most cost-effective approach to improving health or reducing pollution. They shunt polluting activities toward dirtier areas, whereas a redistribution toward cleaner areas would reduce the health and other effects from the same amount of pollutant emission. They add polluters to dirty areas, and may require more stringent and costly emission restrictions in these areas than would have been required to meet standards elsewhere.

The unique feature of nondegradation rules is that they contradict the assumptions behind national ambient air quality standards. If an area meets both primary and secondary standards, then its air is presumed by legislation to do no damage even to the most sensitive people, or to materials, vegetation, and so on. If so, then degradation up to the most stringent secondary standard can bring no harm to the area, and preventing degradation, therefore, can bring no benefit.

8 Energy-Environment Trade-Offs

Various methods of reducing air pollution incur energy penalties. This drawback has become a greater concern with the recent commitment to reduce dependence on imported oil and to economize on energy. Meanwhile, the large increases in the prices of all fuels have magnified the impact of fuel penalties. Last, the emphasis on coal to achieve energy independence has aggravated the conflict between energy and the environment, for coal is the most polluting of fuels. The details of federal energy policy remain to be settled by the Congress and confrontation with environmental interests. But the centerpiece of President Carter's energy plan is a two-thirds increase in the use of coal by 1985. In 1977 the Federal Energy Administration issued orders to eleven power plants and fourteen industrial plants banning the use of oil or gas, and orders for fifteen planned industrial plants requiring construction of coal-burning capability. Federal government policy aside, the sharp increase in the price of oil and uncertainties about its future price or possibly its uninterrupted availability encourage greater use of cheap, abundant, domestic coal.

Although much of the controversy concerning the Clean Air Act Amendments has concentrated on motor vehicle pollution and emission standards, perhaps the electric power industry is more important. It accounts for half of the SO_x and a quarter of the particulate emissions, the two pollutants most severely indicted as hazards to health. Electric power plants, like cars, are found in every sizable concentration of population. The doubling rate of electric power production has been every ten years (it has slowed recently), far more rapid than that of motor vehicles or the output of polluting manufacturing industry. There is no prospect for the emigration of electric power generating plants and the import of electricity. Likewise, the carpark and its increasing use will remain with us indefinitely. On the other hand, many of the polluting manufacturing activities are moving out, or there is a definite prospect that they will move out, to other countries, with increased imports of the finished products. A growing proportion of metallic ores and petroleum products are being imported. The supplier nations are moving toward their own processing facilities, so that the United States will be forced, whether or not it wishes, to import more of the finished products of the metallurgical and petrochemical industries, and conduct less of the polluting processing activities on its own soil.

What are alternative ways of reducing energy-related pollution in power plants and other users of fuel (excluding motor vehicles)? First, the use of clean fuels may be expanded. But the supply of natural gas is already inadequate. (The lifting of price ceilings on interstate sales of natural gas will increase supply). Coal gasification and/or liquefaction would provide an adequate supply of clean domestic fuel, but would take a great deal of time and might prove expensive. The supply of low-sulfur coal is limited (and needed for metallurgical industries): so is the supply of low-sulfur oil, much of which must be imported anyway. Coal desulfurization is experimental, would take time to develop, and might prove quite expensive.

Second, the consumption of fuel may be curtailed. This approach can contribute very little, since in fact what is realistic is a slower increase in use of fuel, rather than any significant absolute decline.

Since the supply of clean fuel is limited for a good many years ahead and there is no realistic prospect of a sizable decline in consumption, the main alternative is the third: stack-cleaning equipment, allowing greater use of domestic coal. Even with stack scrubbers, the planned shift to coal would *increase* SO_2 emissions 5 percent by 1985.[1] A fourth alternative is more rapid expansion of nuclear power capacity. These diverse policies would have different effects regionally on the availability and cost of energy. Regions vary widely in their supply and potential of clean energy sources: hydroelectric power, natural gas, and low-sulfur coal.

"King Coal" and Alternative Energy Sources

Whatever the role of solar energy, wind, magnetro hydrodynamics, fusion, and other nonpolluting or low-pollution sources of energy in the twenty-first century, they will not be making a large contribution by the end of the twentieth. The alternatives in meeting energy demands remain fossil fuels—coal, oil, gas—hydroelectricity, and nuclear fission. Geothermal energy might become important in limited areas. There is no technological fix on the horizon that will circumvent the conflict between energy demands and clean air goals.

Both the Energy Research and Development Agency (ERDA) and the Council on Environmental Quality have estimated that up to 25 percent of United States energy needs could be supplied by the sun by A.D. 2000. These figures are neither forecasts nor predictions, and remain farfetched. The latest optimistic scenario, for example, obtains large figures by defining solar energy broadly enough to include all energy directly or indirectly derived from the sun (in the short run, since fossil fuels are excluded), such as hydroelectric power, windmills, and energy derived from biomass or organic matter. Attainment is further predicated on a large increase in the price of oil and a substantial federal development effort.[2] Federal research and development expenditures on solar energy have been increased some tenfold to $500 million, but still rate as modest rather than impressive.

Hydroelectricity is a small and diminishing share of energy supply, and most of its potential in the United States has already been developed. This, at least, is the conventional view. David Lilienthal, however, asserts that at least one half of the nation's hydroelectric potential is undeveloped. He cites an EPA study estimating undeveloped hydroelectric potential at 113,000 megawatts, contrasted to existing capacity of only 66,000 megawatts.[3] Much of this undeveloped potential consists of dams and reservoirs, some of which generated power long ago but whose generating facilities have been abandoned, some before the age of electricity; others were constructed for other purposes and were never supplied with generating equipment, which could be added fairly readily. Most of these are small, but some are not, and at current and prospective prices for energy they would be economical and pollution-free. Small hydro plants, like some big ones, have problems with imperfect correspondence between their location and the geography of energy needs, and with seasonal flow and inadequate reservoir capacity.

The cleanest energy source at point of use is electricity. How clean it is considering generation, transmission, and final use depends on the source of energy: hydroelectricity is especially clean; low-grade coal is very dirty. It also matters where electricity is generated: whether there is or is not a high concentration of population downwind, and whether the level of air pollution (apart from generation of electricity) is high or low.

The choice of fuels used in households is particularly important for indoor air pollution. Most people spend most of their lives indoors. Electricity clearly is the cleanest, absolutely pollution-free at the point of use. Although gas is the cleanest hydrocarbon fuel, it does pollute, particularly in the confined spaces of kitchen and furnace room. These are the alternatives for cooking and hot-water heating in nearly all cases, but fuel oil is important for furnaces.

The cleanest fuel is natural gas, which is also the second cleanest way of generating electricity. Its problems include its unsuitability as a motor vehicle fuel; the limited known reserves and declining domestic output; and the high cost of imported liquefied natural gas (LNG), or, eventually, of natural gas piped from the North Slope. Prospects for importation by pipeline from Mexico are uncertain at this time. Its use as a source of electricity is a misallocation of a limited supply, resulting from the low price of natural gas when it enters interstate commerce. Rising ceilings on the interstate price of natural gas, and eventual deregulation, provide incentives for exploration and development and for more efficient allocation of existing supplies. It is premature to judge whether the supply response will be large enough to postpone a shift to coal.

Large supplies of gas equivalent to natural gas could be manufactured from our abundant supplies of coal. This could only be done on a scale large enough to reduce air pollution in the longer run, after huge investment in suitable plants. The costs would obviously be much higher than the cost of coal, and the manufacturing process itself involves air pollution, hence the net gain in cleanliness is less than in the case of natural gas. The Council of Environmental

Quality examined four alternative ways of using coal on the border of Wyoming and Montana for space heating in Chicago. Conversion of the coal into high-Btu gas, which was then burned directly to provide heat, proved to be 50 percent more efficient and 40 percent cheaper than conversion into electricity with use of a scrubber. It was also far less polluting.[4]

The difference in cost per Btu between manufactured gas and the coal from which it is made is one measure of the future cost of emission abatement. It is possible to compare the costs of emission abatement by various techniques for coal and oil, since they have the same major air pollutants, although not in the same proportions. The problem of which pollutants to include, and their weighting, inevitably besets any index of emission or pollution abatement.

The relevant price is the marginal cost, which at this writing is the East Coast-delivered price of Algerian liquefied natural gas ($4.50 per 1,000 cubic feet), both for natural gas and its manufactured equivalent. Should there be a restriction of supply of imported LNG, then a higher manufactured price might become the relevant marginal cost.

The remaining alternatives are coal and petroleum. As to petroleum, as in gas, national policy has created a multiple price system: old versus new oil, domestic versus imported oil. The relevant price is the marginal cost of oil, which is the East Coast-delivered price of OPEC oil (realizing that there are differences in composition of oil and in corresponding prices as well as in national price policies).

Clearly, coal is the cheapest fuel to use where it can be employed directly. It is also the dirtiest. It is not yet the lowest marginal-cost fuel when it is liquefied as a source of petroleum products or gasified as a substitute for natural gas. Both petroleum and coal lack uniformity. Different types can vary widely in pollutant emission per Btu generated. The Clean Air Act and similar legislation in many other countries has made low-sulfur coal, and petroleum in particular, scarce relative to demand, so that their price has risen relative to other types of coal and petroleum. Coals vary widely also in other types of emission and in residues per Btu generated.

The Ultimate Pollutants

Combustion pollutes. Fossil fuels contain minute amounts of trace elements which cannot be eliminated by any economic technique. Even the cleanest fuels, methane or alcohol, will produce some pollution in the process of combustion. In the remote (or perhaps not-so-remote) future, humanity will approach the maximum sustainable rate of fuel combustion. Whether problems of pollution will dictate restriction of combustion before this point is reached depends on the progress of antipollution technology. There is, however, one product of combustion that cannot be reduced: carbon dioxide. This is not a health problem, but a long-range ecological threat.

Excessive reliance on combustion, no matter how effective emission control may be, poses indirect threats to health. In the case of coal, the main effects of increased mining (apart from the health of the miners themselves) is acid runoff and pollution of waterways. But much of the world depends on wood and charcoal, whose growing use has denuded vast areas, exposing them to erosion by rain and wind, adding to the particulate burden of the atmosphere, and also reducing the ability of the earth to absorb carbon dioxide from the air. (Carbon dioxide-enhanced photosynthesis will to some extent counteract the effects of deforestation). Atmospheric particulates, whatever their source, in turn modify weather and contribute to a warming trend in climate.[5]

The main product of combustion is carbon dioxide, which is harmless to health. It is the final product of the oxidation of CO, which is harmful in concentrations attained in areas of congested traffic. But there is some concern that the emission of carbon dioxide will eventually increase the atmosphere concentration to such an extent that it will bring about lasting changes in climate through a "greenhouse" effect. The various measures to reduce harmful emissions by using clean fuels or emission reduction devices would not lessen this threat; only energy conservation could. It is estimated that since the beginning of the Industrial Revolution the concentration of carbon dioxide in the atmosphere has risen 13 percent. Current rates of increase in its emission are expected eventually to elevate the temperature of the earth enough to raise sea levels through melting of polar ice, affecting rainfall as well. It has been estimated that doubling CO_2 in the atmosphere will raise average temperatures two to three degrees centigrade, and much more than this in the polar regions.[6] Although there is no certainty about the rate at which increasing combustion will raise the atmospheric content of CO_2, or the precise relation between this concentration, warming trends, and changes in the world climate machine, it is clear that there is some tolerable upper limit and that current rates of combustion are increasing the concentration.

Combustion produces heat directly as well as indirectly through the CO_2 greenhouse effect. Even the most efficient power plants waste more than half of the heat they generate. Nuclear power, although it produces no CO_2, has a large problem of disposing of waste heat. Although heat, unlike CO_2 which increases the retention of heat, is recognized as an environmental pollutant, it, too, is no threat to health. Neither is this an immediate or a national problem, but global in origin and international in policy needs. The time scale involved is long enough that consideration must anticipate the actual problems by decades. Since heat and CO_2 constitute no clear and present danger, they are not visible, and their consequences are subject to considerable disagreement, they present a need for policy goals that are very difficult for representative governments to adopt.

Coal Versus Nuclear Power

The pollution issue in choosing between coal and nuclear power generation is a choice between an evil we know and an evil we ignore. It is also a choice between a certain evil, reducible at known cost, and the remote but not nonexistent risk of nuclear accident, a risk that can be reduced, but not to zero. Fast breeder reactors, already operating experimentally, would increase the hazards because of their generation of plutonium, but not necessarily the (unknown) probabilities of accident. Our concern is not with the numerous industrial safety hazards largely limited to workers in particular types of plants, but with the more widespread radiation exposure resulting from proximity to an operating nuclear plant, and also the exposure that would result from an accidental rupturing or sabotage of vessels containing uranium or plutonium.[7] In time, as more and more radiation-contaminated wastes have to be disposed of, the risks will increase, and the number of areas threatened will multiply.[8] Contamination may be of the air, the water, the land, or all three.

In comparing fossil fuels with nuclear power, quite different emissions must be offset or traded off against one another. It is a question of radioactivity versus sulfates and particulates. So far as radioactivity is concerned, what matters is mainly cumulative exposure; there may be no minimum threshold of health damage, and risks may be independent of the distribution of exposure over time. For emissions from coal and oil power plants, peaks of exposure appear to be more important than cumulative totals, and there are thresholds; the dosage-damage relation is not linear. More important, perhaps, than differences between radioactivity and other pollutant emissions is the differences in risks involved. From coal- and oil-burning plants, the risks are from actual emissions, which are known or readily knowable. From nuclear plants, the main concern is over possible rather than actual emissions—over accidents which could release large amounts of radioactive materials in a short period, rather than in continuous emissions, which are far below any level for serious concern. (Exposure to radiation during the uranium mining process is another problem, and constitutes the largest health hazard associated with nuclear plants. This exposure is not the result of accidents, and it is restricted to the limited number of individuals engaged in mining and handling uranium.)

There is a qualitative difference in the risks involved. Combustion of fossil fuels aggravates asthma, emphysema, bronchitis. Much of the effect is short-run and graduated with the concentration of pollutant. It is rarely fatal, and then mainly among those already suffering from advanced respiratory disease. (It is true that it is also associated with higher incidence of various types of cancer.)

Radiation, on the other hand, is associated only with cancer, particularly leukemia. The public does not recognize such a thing as a mild case of cancer. Death rates for cancer victims are many times higher than for sufferers from respiratory disease. Variations in concentration and in duration of exposure will be associated with variations in the probability of getting cancer, but not with the severity of the cancers that do develop. Furthermore, because of long lags, those once exposed to a great deal of radiation will never feel safe from its consequences. Since the possible results of exposure to radiation are viewed as so very much worse than those from exposure to emissions of coal-fired furnaces, the risk-avoider will prefer the latter even with much higher probabilities of suffering ill effects.

The pollutants released by coal-fired power plants are eventually eliminated from the air. Thus, pollution is related to emissions in the present and recent past (although health damage is related to pollution over considerable periods). On the other hand, the amount of radioactive wastes with a long half-life is for all practical purposes related not to current and recent-past generation of nuclear power, but to cumulated generation of such energy from the very first nuclear plant. The cumulative nature of this hazard therefore causes us to favor coal over nuclear power. However, that nature remains a set of unknown probabilities of various levels and durations of radioactive emission exposure, rather than the much more assured and calculable emissions and exposure to emissions from coal-fired plants.

Electricity versus Direct Combustion

Thermal electric power plants have achieved a plateau of roughly constant costs. There are few scale economies left in thermal electric power generation. Some economies can be achieved from utilization of cheap local fuels (for instance, the low-grade coal resources of the Southwest) which justify long-range transmission of electric energy. Economies in transmission costs can either involve reduced kilowatt miles through better spacing of generating plants and interlocking of transmission networks, or major capital investments in new transmission technology. Investment in transmission amounts to 42 percent of the investment in generation of electric power. No pollution is involved. Thus, the advantages of energy derived from thermal electric power plants, which are few in number and large in size, are not in cost, but in greater feasibility of pollution control and monitoring at the source as well as in lower pollution control costs. These advantages, however, involve much greater consumption of coal than if the coal were used directly for heating at the point of destination.

The conversion of coal into electricity, which in turn will be converted into heat or mechanical energy, is an extremely wasteful process. Even in large modern plants, approximately 60 percent of the heat is lost. There are additional transmission losses. Where the product is heat or mechanical energy, direct generation in situ is often more economical with fuel than initial generation of electricity by burning fuel, transmission, and conversion into the product. This approach toward energy conservation, however, would clearly aggravate air pollution. This results because a large number of small furnaces, stoves, and boilers are more polluting than an electric generating plant using the same amount of fuel, or even generating the same amount of heat or mechanical energy, and because the fuel will be consumed in areas of concentrated population, thus increasing exposure per unit of pollutant emission.

However, if coal is first converted into gas, and the gas burned as fuel, air pollution can be minimized at the point of final use. The conversion of coal into gas is itself polluting, but need not be done near concentrations of population. This would allow millions of sources of fuel consumption with limited air pollution, avoiding the problems of monitoring and pollution control at an increased cost for fuel or energy.

Converting autos and commercial vehicles used for short hauls and/or intermittently to electric propulsion involves an energy penalty, as well as a trade-off between pollutants (SO_2, CO, and particulates versus CO, possibly lead, and NO_x) and different locations of pollution.

Utilizing Waste Heat

Thermal electric power generation is tremendously wasteful. Even very large plants waste more than one kilowatt of heat for every kilowatt of electricity gnerated. Nuclear plants are even more wasteful of heat. We are not concerned here with the thermal pollution problems which result, but with the waste of energy and ways of reducing it. Clearly, the most important single way of economizing on energy is to use the wasted heat. This use might not reduce the need for electricity much, but it would very significantly reduce the need for fuels, mainly oil and gas, which are used for space heating. Waste heat from large industrial boilers as well as from power plants could be used. In Europe, which is much more densely populated than the United States, it is common practice to use waste heat from thermal electric generating plants for district space heating. Such heat would be competitive with imported oil for a large proportion of the U.S. urban population.[9] Such a policy in the United States, however, would involve relocation of the electric power generating industry, with possible adverse effects on air quality and health. It is not that relocation would increase emissions. In fact, as it involves new plants, it would reduce emissions. But the emissions would all be concentrated in major centers of population.

The tendency has been to locate electric power generating plants near the supplies of fuel, which typically means in areas of limited population and pollution, and hence of very small pollution exposure. This was true before environmental quality became a major issue, and was based largely on the fact that the burning of hydrocarbon fuels is a very weight-losing activity and transmission of electric power is much cheaper than shipment of fuel. An additional consideration may now be the possibility of avoiding costs of pollution control in areas with good air quality. Of course, if environmental degradation is not to be allowed anywhere, then there would be no added advantage in locating in areas with clean air. In any case, relocation of the industry toward centers of population and economic activity would mean higher-cost electric power. Siting costs and problems would be much greater (another factor in the trend toward location near fuel supplies) and transport costs for fuel would increase by much more than any reduction in electricity transmission costs.

There are trade-offs, however. Utilization of waste heat for space heating would reduce burning of hydrocarbons for this purpose elsewhere. Since a large electric generating plant can be designed to emit less pollution more easily than a large number of furnaces providing the same amount of heat, it is conceivable, at least in some areas where clean fuels are not available, that a combination of location of power plants near urban areas and utilization of their waste heat would reduce air pollution exposure. More likely, however, in most cases, would be an increase in emissions, particularly when alternative means of heating employ relatively clean fuels. There would be substantial economy in fuel, specifically in the fuels in short supply, natural gas, and increasingly imported petroleum. To the extent that availability of waste heat from thermal power plants preempts the conversion of furnaces from gas or oil to coal, or the installation of new coal-burning furnaces, there would be a net reduction in urban air pollution.

The cogeneration of electricity is the obverse of district heating with the waste heat from electric utilities. Some large industrial boilers could harness the steam they produce for other purposes—to drive turbines and generate electricity—without any additional consumption of fuel. The possibility for combining generation of electricity and process heat is illustrated by West German industry. Twelve percent of electricity generation in West Germany represents cogeneration.[10]

Energy Conservation and Growth

Whereas a shift to coal aggravates pollution problems and most antipollution measures involve an energy penalty, energy conservation reduces pollution. There are no antipollution measures that alleviate the energy problem. Most energy measures other than conservation clearly worsen the environmental problem: they have spillover costs. On the other hand, there are policies and investments which are complementary, which reduce both pollution and energy consumption. This is true of all approaches which reduce the need for energy: better insulation, lighter cars, smaller engines, greater efficiency in combustion, lower thermostats in winter, higher ones in summer. A tax on gasoline reduces both energy consumption and air pollution. However, energy conservation has spillover costs in terms of the economic growth that is foregone. So, of course, do a number of policies intended to reduce pollution: emission reduction devices which reduce productivity, locational policies which prevent investment in productive facilities or shift them to locations or technologies considered second-best in terms of productivity. To the extent that emission control and energy conservation are alternative ways of reducing pollution, their cost-effectiveness can be compared. Both sacrifice some growth, and the question is, which imposes the greater penalty? But pollution reduction is the least of the goals of an energy conservation policy.

If conservation measures that reduce air pollution are mainly responses to the energy problem, then their costs cannot be attributable to pollution reduction. However, the modest stress on conservation in President Carter's energy plan would reduce the annual growth in energy use from 3.5 or 4 percent to 2 percent. It would keep the pollution problem from getting worse as fast, but would not ameliorate it.

Significant conservation measures have been adopted for motor vehicles, requiring an increase in average new-car mileage from 18 miles per gallon in 1977 to 27.5 miles per gallon in 1985. If this improvement is attained, it would result in a net decline in gasoline consumption, even allowing for increase in number of cars and miles driven. The costs, which again are more reasonably attributable to our energy goals than to our environmental goals, would include the costs of research, retooling, speeded-up depreciation for plants and equipment not adaptable to new requirements in manufacturing (which would be passed on to the buyer), plus some loss of consumer satisfaction associated with large, powerful cars. The spotty record of Detroit's efforts to market truly compact, economical cars time and again testifies to the strength of consumer preference for size, weight, power, acceleration, speed. But of course consumers also prefer clean air, even if they neither receive nor perceive health benefits for themselves. It is not clear that there would be a net welfare loss to consumers, since the preference for large, powerful cars may be a relative one, and a reduction in size and weight of all cars might leave consumers in their accustomed or preferred relative position, with lower costs of driving a car. (Weight reduction by economizing on materials indirectly further economizes on energy.)

On the other hand, air quality legislation and regulation impose a constraint on energy programs. If air quality could be ignored, the planned shift to coal from oil and gas could be accelerated, made more radical, and much cheaper, for it would not have to be accompanied by expensive measures, such as smokestack scrubbers for SO_2, to hold down the amount of emissions. Thus, one of the costs of air pollution restrictions is the constraint imposed on conversion to coal.

There is a further question: How will the higher price of energy affect economic growth? (Most of the increase is attributable to OPEC rather than to environmental legislation.) A large part of energy consumption—electricity, gas, fuel oil, gasoline—is by households. Household use in turn is in part for household production—cooking, washing, heating and cooling, commuting—and in part for final consumption. The distinction between the two is a complex conceptual and practical problem which need not concern us here. Household production is not included in the national income accounts, although it is certainly a significant component of real output and an employer of labor and resources. Energy consumption by business, on the other hand, is included as part of the process of production and is directly reflected in costs and output.

Is there much waste of energy which could be eliminated without affecting production? We know that energy consumption per capita in the United States is almost twice what it is in West Germany or Sweden. Although overheating in the winter could be eliminated at no cost, as could the use of overpowered cars for urban commuting, most economizing of energy or elimination of waste is costly. That is why little has been done here by contrast with other advanced countries. The cost of energy in the United States has been much lower than in West Germany or Sweden.[11] Energy conservation takes labor, capital, and managerial attention. How much is done depends upon the price of the energy conserved. Responsiveness to sharp price increases in energy in recent years is already evident in a preliminary way (some adjustments involve investments with long lead times). Consumption of energy in the United States rose only 1.4 percent between 1973 and 1977, whereas real GNP rose over 8 percent. In the immediately preceding years, energy use had increased in proportion to GNP. The industrial sector, most keenly aware of its energy costs, reduced consumption 13.5 percent between 1973 and 1976.

The main differences in energy utilization between the United States and Western Europe, apart from overheating of space and the unavoidably greater use of transportation associated with large area and low population density, are associated with urban form and commuting patterns.[12] Urban spread, the prevalence of single-family, detached housing, and dependence on automobiles for commuting are all mutually interdependent, all energy-wasteful, and all to some extent the consequence of low prices for gasoline and for auto transportation in general. Thus, measures to reduce the use of cars have widespread effects on energy usage in the long run, beyond the immediate saving in gasoline. Measures limited to improving car efficiency (greater gasoline mileage), however, have the opposite effects; although they may economize on gasoline, they encourage detached housing and suburbanization.

Low pollutant fuel is considerably more costly per Btu than high pollutant fuel (scarce natural gas excepted). To the extent that this differential is a scarcity rent paid within the nation (the case with low-sulfur coal), it amounts to some internal income redistribution. To the extent that the clean fuels are imported (the case with petroleum), the income redistribution is a payment to foreigners and a net loss to the nation. In both cases fuel as an input becomes more expensive relative to everything else.

The higher price of fuels, then, has two consequences: it reduces the domestic income and output insofar as it involves transfer of resources abroad, and alters or distorts the process of production by making one input much more expensive than before. A study by Karnovsky of the St. Louis Federal Reserve Bank claims that the combination of environmental safety costs and increased oil costs had decreased the real economic capacity of the nation by 4.5 percent at the end of 1973.

9 Policy Alternatives

In retrospect, the EPA has shown commendable flexibility. It has postponed the 1975-76 deadlines for emissions and for ambient air quality. It has reconsidered its standard for NO_x. It has recognized the impossibility of arriving at technical solutions on schedule. But some damage has been done. The pressure of time imposed by legislation has committed the auto industry to emission reduction techniques that could reasonably be developed in a short time, rather than to techniques that might prove optimal in the longer run. The same may now be true of coal-burning utilities. This means that the eventual cost will be greater, and that the day on which better approaches will be incorporated in new car models has been postponed for years.

Legislators, and the EPA, had to make decisions in considerable ignorance. They realized that the necessary technology did not exist in the case of motor vehicles. Some believed that technology would not be developed except under the threat of execution which the 1975-76 deadline provided. Some also believed that with this incentive the industry would promptly develop emission control techniques and incorporate them in millions of new cars. This proved not to be the case. The faith in the omnipotence of science and technology proved misplaced, in the time frame allowed. The problem was to assure that the manufacturers put on their best efforts. Decisions had to be made in the presence of uncertainty as to the progress of emission abatement technology, the relation between emissions, air quality, and pollution exposure, and the effects of pollution exposure on health. All relevant facts are never in; the optimal technology is rarely on hand. The deadline was a bluff that had to be modified in the light of what could be reasonably achieved. The approach selected was regulation to achieve specified emission and air quality standards. Are there approaches toward achieving cleaner air that are more cost-effective, or offer greater returns, than the regulatory approach toward emission control? As to regulation itself, what alternatives may be suggested?

Who Should Regulate What, and How?

One way of minimizing the cost of achieving primary air quality standards everywhere is to vary emission standards by area. Areas with relatively pure air would be excluded from control; the dirtier the air, the stricter the emission rules would be. As the number of emission sources increases, stricter standards would be enforced and more areas would be regulated. This approach, however, is not practicable for many sources of emission, particularly for emission control that is part of the design of products (such as motor vehicle engines) or that applies to mobile sources.

Although there is a single federal emission standard for all new motor vehicles of a given type, standards for different types of vehicles differ widely in stringency, a variation not justified by differences in costs or benefits. In 1975, emissions should have been reduced to 17 percent of their uncontrolled level for automobiles, 25 percent for light trucks, but only 47 percent for heavy gasoline-engine trucks and 75 percent for heavy diesels. Lawrence J. White estimated that as a result, total emissions from trucks and buses were becoming nearly as great as emissions from cars.[1]

The ideal approach is to leave each area free to choose the most cost-effective way of achieving clean air given its particular combination of pollutants and emission sources. It is not feasible to build different emission characteristics into motor vehicles to be sold in different metropolitan areas, partly because of the increased cost of multiple standards, partly because of the mobility of Americans. But a two-car strategy, with low-emission cars sold in polluted areas and medium-emission cars sold elsewhere, is feasible. In effect there is a two-car strategy, because California has established emission standards stricter than those of the federal government. But neither in California nor in the other states is there a two-car policy geared to differences in local or regional conditions. There would be enforcement costs at annual inspection, particularly if there is a substantial difference in price and/or operating costs between low- and medium-pollution cars. There would be a penalty on migrants from dirty to clean areas who have invested in a more expensive low-pollution car.

Even with a one-car strategy, emission standards are not completely built in. Localities can adjust to their particular trade-offs by varying the frequency and standards of inspection. They may also require retrofitting of old vehicles, instead of limiting enforcement of emission standards to new cars. In addition to emission inspection and retrofitting, local policies may include mandatory maintenance, traffic flow measures, traffic rescheduling to off-peak hours, and conversion of high-use vehicles such as taxis, trucks, and buses to nonpolluting fuels. Zerbe finds local strategies much less costly than national new-car standards, but equally effective.[2] Grad and others conclude that emission inspection is the most cost-effective strategy of all.[3]

The same considerations apply to stationary sources: emission standards might vary by area, retrofitting might or might not be required, enforcement might be strict or lenient. The range of pollutant emission sources regulated might vary. An industrial center should adopt a different regulatory mix and annual emission threshold from those appropriate for a trade and service center.

Ideally, polluters would be free to choose the most cost-effective method of meeting emission standards. The requirement that all new public utilities burning coal use the best available technology to reduce emissions—namely, stack scrubbers—will discourage use of expensive low-sulfur coal as well as development or adoption of alternative technologies, such as removal of sulfur and pyrites from coal before burning (or tall stacks, where airborne concentration of pollutants rather than emissions is the criterion). A policy setting emission limits and leaving it to each public utility how best to meet them would encourage promising lines of research and would not penalize adoption of other techniques; it would be neutral between alternative means toward the same end. Such a policy would require more monitoring, but it cannot be assumed that the compulsory installation of stack scrubbers avoids the monitoring problem. On the contrary, the efficiency of stack scrubbers is highly dependent on frequent and expensive maintenance. Without regular monitoring, many utilities may opt to neglect this maintenance, with a rapid decline in emission reduction efficiency.

These approaches yield the least-cost attainment of national ambient air quality standards. They do not assure attainment or maintenance of higher air quality in areas that already meet the national standard. Nor do they equate the incremental cost of pollution abatement with the incremental benefits in any particular region, except by accident. This last objective would argue for different regional standards rather than a single national one, or for other types of policy. For the immediate future, the necessary information on the marginal benefits from abatement to determine optimal air quality standards on a regional basis does not exist.

Polluted air spills over city, county, and even state boundaries. Its flow is less certain than that of polluted water. Polluting localities have inadequate incentives to limit pollution, since they are exporting part of the pollutants they generate. They also have limited incentives to act unilaterally, for two reasons: they will continue to be affected by pollution generated by their neighbors, and local costs of pollution abatement will hamper them in competition with localities that do not incur such costs. Localities receiving polluted air from elsewhere have no power to deal with the problem directly.

The federal government has taken the central role in reducing air pollution, although we should not overlook the fact that some communities with a severe, well-recognized problem, such as Pittsburgh and St. Louis, took effective action on their own long before the Clean Air Act.

Of the ubiquitous sources of air pollution, electric power plants are eminently suited for pollution prevention and monitoring at the source. Individual cars, on the other hand, cannot be monitored constantly. Periodic inspections and maintenance would require expanded facilities and training of mechanics. Whether substitutes for monitoring or universal inspection, such as sample checks with heavy penalties for substandard performance, could achieve the same results, possibly at lower cost, is an open question. Political and psychological acceptability is also an issue. Motor vehicle pollution can be regulated via the manufacturer, by requiring engine modifications or postcombustion devices, and via the fuel supplier, by regulating gasoline constituents. Most stationary sources of air pollution (mainly residential and commercial heating and cooling plants) are not easily controllable at the source since there are nearly as many as there are cars, nor in the short run can they be controlled through regulation of manufacturers since the turnover of furnaces and associated equipment is much slower than for motor vehicles. Only regulation of fuel composition is feasible in the short run. This technique may also be the most cost-effective in the long run.

Thus, the choice of pollutants to be regulated and sites of regulation has reflected feasibility and cost rather than contribution to the total air pollution problem. Residential and commercial stationary sources have been ignored. Motor vehicles have been tackled through manufacturers, while power plants and industrial emitters have been tackled directly.

Policy Options

In devising an optimal pollution abatement strategy, several questions must be asked. First, what is the effect of air pollution on health? Second, is it feasible to reduce air pollution everywhere sufficiently to minimize its adverse effects on health? Wherever it is not feasible, then what policies for minimizing exposure to air pollution, or for reducing sensitivity to the harmful effects, or for treating the adverse effects, are most cost-effective? Even if it is feasible to reduce air pollution sufficiently to minimize its adverse effects on health, it is still pertinent to ask which type of policy (or which policy mix) is most cost-effective: reduction of air pollution, reduction of exposure to pollution, reduction of sensitivity (e.g., allergy shots), or treatment of the adverse effects. The answer may change from time to time as a result of technological advances. It may be different depending on the time period over which health hazards are minimized, because some policies can be instituted promptly and others take time.

There is a wide spectrum of conceivable measures for dealing with the air pollution problem. They include the following:

1. Compensation of victims;
2. Treatment of victims;
3. Minimization of exposure;
4. Reduction of contamination ex post facto;
5. Elimination of contaminants at the emission source;
6. Prevention of contamination.

Pollution is a by-product of production and consumption. In other words, there are benefits associated with pollution. When the potential damage is less than the associated benefits, the optimal policy could be number 1 or 2 above. If damage is defined as the amount required to compensate victims, then compensation is a conceivable policy. Whether treatment is feasible or optimal depends on the state of the medical arts and the costs of treating victims. Policies 3, 4, 5, and 6 are all feasible to some degree, at some cost. The choice between them will depend on the state of pollution control technology, on the availability of nonpolluting substitutes, and on the cost of reducing exposure. When the potential damage is greater than the associated benefits of polluting production or consumption, then the optimal policy must include number 3, 4, 5, or 6.

Compensation can be excluded as a major policy, because of the impossibility of identifying most of the victims and measuring damage to their health and its value. (This is not the case in numerous instances of highly localized industrial pollution.) Compensation runs the risk of weakening incentives for potential victims to minimize exposure and for the public to minimize emissions. Polluters and victims are largely the same people. Also, this approach runs counter to American values, which seek to solve problems by eliminating them.

Treatment likewise is not an acceptable substitute for other policies, because many victims cannot be identified, their conditions cannot always be treated successfully, and the American people will not accept treatment as an adequate response to air pollution.

Minimization of exposure is a feasible, although limited, policy. It includes separation of emission sources from the population, controlled air quality in confined areas (indoor air filtering), and individual protection through masks. Much emission cannot be separated from the population, and people are not willing to wear gas masks or stay indoors most of the time. This is not to say that these policies have nothing to contribute; only that the main reliance must be on emission control rather than exposure control, treatment, or compensation.

Consideration of outdoor air pollution assumes that indoor pollution is adequately indicated by measures of outdoor pollution or that it poses no health problem, or it overlooks the fact that most people spend more of their lives indoors than out. The difference in pollution can be large. The relative feasibility and cost of alternative policies for coping with health hazards indoors are quite different from those needed outdoors. In a contained environment, air pollution is more easily managed. It is possible to filter many outdoor pollutants from the indoor atmosphere; indoor sources can be eliminated. On the other hand, minimizing exposure is not feasible; in fact, the major way of minimizing exposure is by staying indoors and avoiding outdoor air pollution.

The idea of treating the air may appear farfetched, but in fact it is nothing new in the area of weather control. Cloud seeding to induce rain is well known, sometimes successful, and widely practiced. Smoke and man-made mists that prevent freezing are standard practices in Florida and California. Attempts have been made by the Air Force to divert incipient hurricane formations by bombing. Pollution is regularly reduced or eliminated by natural processes: wind, which disperses it; rain, which precipitates it; deposition through gravitational pull; chemical changes in the air which result in precipitation and absorption. The natural processes simply do not work fast enough in areas of high man-made emissions to limit pollutant concentrations to acceptable levels. Techniques for accelerating or supplementing natural processes therefore should logically be considered.

It has been suggested that some pollutants—PAN and ozone in particular—can be temporarily eliminated from the air by chemical means. Chemicals such as diethylhydroxylamine (DEHA) could be sprayed when pollution is high, or when a dangerous buildup is foreseen. The cost could be a small fraction of the cost of emission controls. PAN and ozone are not emitted, but are the result of chemical changes in the air, to which emitted pollutants such as NO_x do contribute. (A feasibility test has been proposed for Jerusalem.) A major question is whether the chemicals used to clean the air may not themselves cause damage to health after repeated use over long periods.[4] This approach, although untested, offers the intriguing possibility of coping with pollutants of natural origin that are unaffected by legislation.

Postemission air quality control is currently only feasible in confined spaces; hence, the central policy choice lies between alternative ways of reducing emissions of pollutants (such as sulfur scrubbing equipment) or preventing the use of polluting materials (switching from coal to gas).

Policy Instruments

There are three main policy tools that could be employed to deal with the problem of air pollution. They are (a) taxes or charges, (b) subsidies, and (c) regulation, of which emission and air quality standards are only two among numerous options. Two or more of these instruments may be combined to implement a policy. The major distinction between them is their treatment of the right to pollute the air. A tax or charge on pollution emission sets a price on polluting the environment, which is levied on the polluter, who is also the beneficiary of the use of the air to dispose of wastes. Regulation, by forbidding emissions, requiring emission-reduction equipment, or constraining behavior (for example, altering traffic flow or requiring cutbacks in polluting output during air pollution alerts), also imposes costs on the polluter, which, however, may be little related to damage or benefit of waste disposal. Subsidies for those who reduce their emissions least of all resemble benefit charges. In terms of the optimal allocation of the right to pollute within an overall ambient air quality standard, policies imposing costs closely related to pollution emission appear most efficient. Emission standards for each polluter are not the way of obtaining a given emission reduction at least cost, or of obtaining the maximum reduction possible for a given cost. They do not distinguish between polluters in terms of the cost of reducing emissions or in terms of the level of ambient air quality. And they provide no incentive to reduce emissions below the standard. Ambient air quality standards themselves of course provide incentives to reduce pollution to the standard, but none to lower it further.

The major policy tool invoked to date is standard-setting and regulation. Other tools, such as taxation, subsidies, and additional regulations, have been proposed. Standard-setting is necessary for regulation, but not for taxation or subsidy. For taxes and charges, it is necessary to have some measure of pollution emission. For subsidies, some measure of reduction in pollution is necessary if the subsidy is geared to it; if the subsidy is for investment in antipollution equipment, no measurement is necessary.

Tax Policies

One main approach to controlling air pollution is a tax to correct the misallocation of resources that arises when social and private costs diverge. This tax would be a "price" charged the polluter, reflecting the damage inflicted on others. The tax would provide an incentive to the firm or individual to reduce his tax by lowering the level of pollution.

Baumol has suggested that in cases where quantification of the externality is not practicable, the tax rate should be set to achieve standards or levels of pollution that are deemed acceptable.[5] An experimental process that tightens standards whenever they prove to be easily achieved may well converge toward an optimum standard and tax. Businesses aware of the possibility of future changes in a standard would construct what George Stigler refers to as "flexible" plants, designed for future modification at minimum cost in response to changes in standards.

Since the relation between emission standards and air quality standards varies widely from place to place as a function of the concentration of polluters, and as the relation will change over time in most areas, either air quality or emission standards should be flexible. Since the objective is air quality, emission standards should be flexible. This argues for production processes and/or emissions technology that can be varied to modify the rate of emissions as conditions warrant. An emission tax encourages development of more efficient technology, whereas an emission standard does not; there is no benefit in overcompliance. Furthermore, whereas a standard results in choice of an inflexible technology, an emissions tax will lead to choice of a flexible technology, so that the polluter may exercise his option to optimize the amount of pollution, given the tax and his abatement costs.[6]

A tax may have as its objective the reduction of pollution. Or taxation may be used to provide revenue earmarked to reduce pollution through research or subsidy of antipollution measures. A subsidy-like effect can be achieved by taxing on the basis of the amount of pollutant emission. If emission is reduced for whatever reason, then tax liability is also reduced.

The Emission Tax. A tax related to the amount of effluent is widely regarded as the ideal policy to deal with externalities such as air pollution.[7] An emission tax would accelerate the replacement of old equipment, because old equipment is more polluting than new and hence would generate a higher tax burden. The great advantage of an emission tax is that it leaves the producer free to choose the technique that is best for him, whether it be a change of fuel, redesigned equipment, or addition of pollutant-removal devices. He would also be free to decide how much to reduce pollutant emissions, based on his particular situation. The heaviest polluters would be under the greatest pressure to reduce emissions, whereas others might prefer to pay the tax if their emissions and the corresponding tax are small and the cost of reducing them is large. The emissions tax is a least-cost way of reducing pollution. The problem with such a tax is that it may have little relation to local or regional levels of air pollution and therefore to the benefits of abatement, unless rates vary regionally.

Richard Wilson suggests using the "integrated pollution exposure," or the harm done by pollution, rather than the amount of pollution.[8] This approach implies higher tax rates in areas with large concentrations of people and lower rates in sparsely populated areas. It also implies higher tax rates in areas with high levels of ambient air pollution than in areas with low ambient air pollution, for risk is a product of the number of individuals exposed and the level of pollution to which they are exposed. The problems with this approach are practical. Tax rates would have to be set for each defined area. There would be boundary problems. The population at risk would not change rapidly, so that tax rates would not have to be altered frequently on that account. But the level of pollution can change radically in relatively short periods, and does fluctuate widely over time. Such a tax would have to ignore short-run fluctuations in ambient air pollution, leaving to regulation the task of coping with air pollution emergencies. From a cost-benefit point of view, there is no justification for any tax in regions where air quality is good and population sparse. The question is, at what point should government start imposing a tax in regions with a substantial population, economic activity, and rising levels of pollution—should it wait until air quality standards are threatened, or not?

The main difference between an emission tax and a tax on pollution exposure is that the former can be invariant over space, and hence need not affect the location of economic activities. In either case the polluter may affect his tax through emission reduction. In the case of pollution exposure as the tax rate base, the community, and to a limited extent the major polluters in the area, may also affect the tax rate through emission reduction or increase. Thus, taxing on the basis of pollution exposure rather than emission alone gives an area and its major polluters an additional incentive to reduce pollution. Should an emission tax rate be allowed to vary between areas in accordance with their air quality (whether or not the population exposed is taken into account), then the additional incentive to reduce pollution would also be operative. Naturally, the higher the tax rates, the more effective the incentives to curtail emissions. Pollution, however, may be imported or exported, in which case any incentive to reduce pollution is improperly allocated between areas.

Emission taxes may be viewed simply as instruments to achieve standards. Another view is that they are alternative and superior to standards, being far more closely aligned wih the benefits of pollution abatement, because there is no incentive under air quality or emission standards to reduce pollution below the levels allowed by the standard. Crocker points out that imposing standards on emitters alone places a minimum but no maximum economic value on the life- and property-supporting capacity of the air, and a maximum but no minimum value on the waste disposal capacity of the air.[9] A tax on emission alone is a better incentive for the polluter to reduce emissions in the most efficient way and to the optimal extent: that at which the additional costs of emission reduction are just equal to the tax saving. Higher emission tax rates could be imposed on more polluted than on less polluted areas, without undermining incentives for polluters to take action.

Pollution Rights. J.H. Dales has proposed a market in pollution rights.[10] The environmental quality control authority would determine for each defined region the amount of emissions it will allow from controlled sources. This fixed quantum will be based on desired air quality and assumptions of relations between emissions and ambient air quality. (In fact, Dale's proposal was illustrated with water pollution rights, but there is no conceptual difference with air rights, and the main practical difference is the separability of different bodies of water compared to the relatively amorphous and unstable nature of air flows.) Air quality might be disaggregated into the key local pollutants, weighted by an appropriate severity factor, and pollution rights could conceivably be sold for each category of pollutant. These rights would be sold to any willing buyer, who in turn would be free to resell. The environmental control authority (or environmentalists, for that matter) might at some future time reduce the amount of pollution rights by buying up some and "sterilizing" them. Improvements in emission control technology would lower the price of pollution rights; expansion of regulated polluting activities would tend to raise the price.

Auction of pollution rights, like an emission tax, requires adequate monitoring. The difference is that it sets a quantitative limit on the amount of regulated pollutant emissions in an area, which neither emission standards nor emission taxes do. It resembles a quota, whereas a tax resembles a tariff. Monitoring requirements may be less stringent than for emission taxes. It is enough to know that a polluter is not exceeding the emission for which he has purchased a right; the amount of shortfall is irrelevant, except to the polluter himself. If the polluter is exceeding his rights, it is not necessary to know the exact amount, as long as the penalties for unauthorized pollution are substantial.

Pollution rights and emission taxes will result in somewhat different allocations of pollution emissions. Different purchasers will pay different prices for the right to pollute, as prices change and fluctuate, whereas the tax per unit of emission will be the same for all and will remain constant over some time period. Pollution rights will have to be valid over years in many cases, since plant and equipment investment decisions will be based on them. Dales suggests rights with time limits of one to five years. Both pollution rights and emission taxes contain analogous incentives to improve performance, in one case by reducing tax liability, in the other by allowing sale of excess pollution rights. However, the gain is less certain for the seller of rights, since he is not sure of selling at the same price that he paid, although there is the possibility of selling at a higher price. With imperfect capital markets, the barrier to entry of a new polluter is greater when he is faced with the necessity of purchasing pollution rights for a possibly lengthy time in advance of initiating production than when he is faced simply with periodic payment of an emission tax based on previous productive activity.

The advantage of a pollution rights approach is closer control over total emissions and therefore over air quality. As emissions decline, pollution rights are easier to obtain and fall in price. As emissions (including emissions from unregulated sources) rise, the supply of emission rights for sale can be reduced and their price rises. The market itself makes the adjustments in price, whereas in the case of emission taxes, a government authority would periodically have to make specific decisions on altering tax rates. A continuously changing price causes less disruption than infrequent, possibly large changes in tax rates.

Fuel Taxes. The emission tax approach is appropriate for electric power plants and many stationary sources. On the other hand, the monitoring required for an emission tax cannot be considered for individual motor vehicles, because with over one hundred million of them, the cost would be prohibitive. If mileage is used as the basis for determining the amount of emissions, there must be periodic inspection of the odometer. What are the chances of cheating? What effect does a tax that is paid at infrequent intervals have in reducing car use? Since the harm from and the amount of pollutants emitted by a car depend on the conditions under which the car is driven (acceleration, stop-starts, traffic congestion), a mileage tax does not adequately reflect emissions.

A tax can be based not on mileage but on fuel consumed. An advantage of a tax on fuel over one on mileage is that it is paid more often, hence the driver is likely to be more sensitive to it, seeing more clearly the relation between his decisions and his tax liability. The administrative cost of a tax collected from fuel suppliers is obviously much lower than that of one on mileage, which must be assessed periodically and collected from car owners. A further advantage is that a tax on fuel consumption is a closer approximation to pollutant emission than a tax on mileage. Large cars, inefficient cars, and cars driven in congested areas consume more gas for the same mileage.

Nevertheless, a tax on fuel is a good indicator of emissions only in the short run, with little change in engine performance. It is not a very good surrogate for an emission tax in the long run because most of the harmful emissions are the result of engine design and performance rather than of the amount of fuel consumed. For a given engine design, however, it becomes a closer approximation to an emission tax and if regular inspection and maintenance are required, it comes closer still.

Ownership and operation of a private car are cheap in the United States, in terms of the purchase price and cost of gasoline relative to income. Thus, increasing the marginal cost of car use is a very difficult way of reducing emissions.[11] While it is true that if a tax is high enough, it would be in the self-interest of individual owners to reduce usage, opt for cars with lower costs of operation, or shift to some extent to other forms of transportation, the question is, How high does a tax have to be before it significantly modifies the behavior of the average car user?

In the short run, the elasticity of demand for car use relative to the price of fuel is close to zero over a wide range of prices. The tax necessary to attain given amounts of pollution abatement is far in excess of any estimated damage cost of driving, which Zerbe has put at one cent per gallon of gasoline consumed.[12] Only in the long run, as decisions are made on what type of car to purchase and what transportation modes to use, is there much response. Even in the long run, spatial patterns of residence, employment, and services are such that many find a car the only feasible or the most convenient form of transportation. The problem is partly that much of the marginal cost of using a car is not paid at frequent intervals, but in infrequent lump sums—annual license taxes, annual insurance premiums—so that the user does not consider full marginal cost. Changes in method of payment that convert such lump sums into frequent charges closely associated with car usage might bring about a more significant short-run response.

There are other negative externalities of car use in large cities that justify a tax much higher than that necessary to equal the damage caused by exhaust emissions. Caution is in order, however, considering the fact that in Europe gasoline costs three or four times what it costs in the United States, but nevertheless, most European countries have experienced very rapid growth in car ownership and usage as incomes have risen.

While a tax on gasoline may not be the perfect surrogate for an emission tax, it may well serve a purpose in discouraging some use of cars. Taxing gasoline consumption can also be viewed as taxing the consumer of the product that causes pollution. It may be a way of achieving some realignment of the cost of owning and driving a car to include the cost of externalities. While the tax may not cause a marked shift in demand, at least it would be a source of revenues to finance air pollution control programs.

Taxes on fuel composition have also been proposed. A tax on the lead content of fuel would be appropriate and feasible, because it would be collected from a limited number of producers.[13] The Pure Air Tax Act of 1972, proposed by President Nixon to the Congress, would levy a tax of ten to fifteen cents per pound of sulfur in fuel or per pound of sulfur emitted into the atmosphere in regions failing to meet the national primary and secondary air quality standards for sulfur oxides, beginning with 1976.[14] The tax rate for each year would be based on a region's air quality in the preceding year. The sulfur tax would be expected to stimulate polluters to develop and install control technology and to use clean fuels as quickly as possible in order to minimize their tax liability.

There is no doubt about the merit of taxing large sources on the basis of their sulfur emissions. But the cost of the monitoring necessary to tax small sources would far exceed the benefits. Appropriately, the proposed legislation would tax small emitters according to the sulfur content of the fuel they use. No taxes would be imposed on fuel sold for use in motor vehicles, since the amount of sulfur in gasoline is minor compared to that in other fuels.

Taxes on fuel consumption and composition are inferior in principle to an emissions charge because they bias taxpayer responses toward fuel economy or cleaner fuel, even when improved combustion or postcombustion control devices might be superior in performance or more cost-effective in reducing pollution. However, in cases where neither emission monitoring nor emission control standards are practicable, fuel taxes may be a desirable way of reducing emissions from otherwise unregulated sources, so that the entire burden of attaining air quality standards may not be imposed solely on sources subject to emission standards (or taxes or charges).

Other Taxes. If gasoline consumption is not a very good indicator of pollution emission, neither is engine horsepower. Engine performance is the appropriate indicator. If emission control is a permanent operational characteristic of engine design, then an excise tax can be imposed on that basis. Such a tax, however, would be related to engine use only in the aggregate, providing no incentive for the individual to economize on the use of a car once the purchase is made. If emission controls deteriorate, a tax on engine performance characteristics would have to be supplemented with or supplanted by periodic inspections, with fines for poor performance and mandatory maintenance. This is equally true of all other surrogates for an emission tax or charge.

It has been suggested that motor vehicles be required to pay a toll when entering central cities that face severe pollution problems. This suggestion has been refined to limit tolls to rush hours, or to cars with only one rider entering the central city during normal business hours. The intent is to reduce the use of cars, particularly during peak hours, and to encourage car-pooling arrangements, resulting in a reduction of the generation of pollution in places where and at times when emissions are at a maximum. Downtown all-day parking taxes are another method of achieving the same goals.

Institution of a toll system reduces the volume of single-passenger private cars and relieves congestion on the road. Assignment of a special bus lane could allow buses to move faster and reduce commuting time, inducing more passengers to use them while reducing the incentive to drive individual cars. As the volume of bus passengers increased, more buses would be needed. Operation of the transit system might eventually become economically efficient, and with improved service a subsidy would no longer be needed. Although the response of commuters to increases in the cost of commuting are known to be inelastic, a daily charge levied at a strategic time is likely to be more effective than the same amount paid for fuel. It is downtown driving during rush hours that most contributes to peaks in air pollution. A charge related to these peaks is better in terms of allocation than one that is applied indiscriminately to all motor vehicle usage. However, a toll high enough to reduce traffic substantially may be politically infeasible. It would certainly exceed health damage costs.

Subsidies

Subsidies have been recommended for a number of reasons, among them compensation to damaged parties as an incentive to reduce and to seek better means of reducing pollution.

The principal objection to subsidies is that they do not realign costs to present a true picture of the social costs of producing and consuming a product. Subsidies would not correct the distortion in the allocation of resources that results from the exclusion of the costs of air pollution from the cost of the product. They are compatible with output that is too high in polluting industries, and price that is too low. The producer who requires a subsidy in order to operate once he is charged the full social costs of his business is not efficient enough, perhaps, and should be allowed to go out of business. Subsidies may also create a moral hazard; they may provide an incentive to increase pollution in order to increase the likelihood of a subsidy or its size.

Another significant objection to subsidies is that they could stunt technological progress by acting as disincentives to the search for more efficient methods. Except for aids specifically granted for technological research, continuing grants, subsidies, or tax credits given to an industry that is not competitive could make it dependent on continued aid. Another reservation about grants, especially for equipment, is that they may delay action, as firms wait in line for assistance before implementing pollution control techniques. Such a delay can be avoided by using a subsidy in conjunction with other policy instruments, such as an emissions standard or tax that takes effect whether subsidies are approved or not.

Compensation. The use of a subsidy as compensation to damaged parties is not a policy designed to reduce the level of air pollution. It may still be recommended, but strictly on equity grounds. If financed by a tax on polluters, in particular an emission tax the rate of which is such that tax revenue is equal to the damage caused by pollution, it may both provide compensation and reduce the amount of pollution and of required compensation (if abatement costs are less than damage costs). But is is the tax, not the subsidy, that reduces pollution.

Once a tax or charge is imposed on polluters, compensation to damaged parties may have perverse effects. It reduces their reluctance to be exposed to pollution, possibly increasing the total amount of health damage from a given amount of emission and creating another kind of moral hazard. If emissions are not reduced by an emission tax or standard, then both the amount of compensation required and the size of the possible perverse effects are increased. There is a strong equity case for compensation to injured parties who were residents in an area before the introduction of damaging pollution, but not to those who entered the area later. To compensate the latter would preclude arriving at optimal locational decisions. It may be desirable to discourage movement into polluted areas, whether by zoning or fiscal means, as a way of reducing pollution exposure.

There would be no such objection to a subsidy devoted to reducing exposure to pollution. This has precedents in acoustic insulation of buildings near airports, and in industrial activities that expose workers to dangerous pollutants, but no proposal has been made to reduce exposure in dirty cities by subsidizing filters to improve the quality of indoor air. One may ask why the victim needs a subsidy. It may be a question of equal access to better health for all income groups. Or it may be compensation for ignorance or undervaluation of the benefits of reduced exposure.

Equipment Tax Credits. Equipment tax credits appear to be the most popular on tax write-offs and tax credits for the purchase and installation of air pollution equipment.

The U.S. Tax Reform Act of 1969 provides for a five-year straight-line depreciation of certified air pollution control facilities. Accelerated depreciation allows tax savings in the earlier years, reducing the net cost of antipollution equipment. Capital investments on equipment such as taller smokestacks that merely diffuses pollutants are excluded. (Revenues from sale of by-products are subtracted from the value of the equipment eligible for depreciation.) For equipment with a useful life of over fifteen years, only that proportion of total costs normally depreciated over fifteen years would be depreciated over five years.[15]

There are problems associated with tax credits and accelerated write-offs. It may be difficult from an accounting standpoint to allocate costs to pollution control when investment in process changes adopted primarily to increase productivity also reduces air pollution. Tax credits distort the tax structure, and therefore the cost structure of the firm. They promote use of capital equipment and pollution control facilities, while providing no incentive to look for other approaches to air pollution control, such as low-pollutant fuels, at lower costs. If tax write-offs or credits are not specific, they become general subsidies of business. If they are too specific, they limit the choice of alternatives. If external costs are to be made part of production and consumption costs, there appears to be no justification for a subsidy, unless it is to encourage controls that are not required by regulation, or to protect the business from unregulated competitors.

Award Payments. Award payments to help offset costs of reducing emissions of pollutants combine the concepts of subsidies and regulation. These payments could be related to the amount of abatement achieved (that is, the pounds of pollutants recovered), or to a particular standard of performance (for instance, a power plant reaching a 95 percent reduction of potential sulfur oxide emissions). Payments could be adjusted to area conditions. The intent is to induce firms to switch to low-sulfur fuels or to curtail operations in times of critical pollution levels by compensating them for increased costs or losses resulting from such shifts.

Subsidies to Curtail Use of Emission Sources. Subsidies to car pools have been suggested. The possible economies in fuel and associated reduction of emissions and traffic congestion can be very large. An increase of one rider in every commuting car has been estimated to save almost 800,000 barrels of gasoline a day. There might be some leakage, cars no longer needed for commuting would become available for other uses during the day.[16] Such subsidies are not an ideal approach, since they bias choices between automobiles and other, less polluting forms of transportation, but given the low elasticities associated with demand for mass transit, this effect may be negligible. They do improve the efficiency with which automobiles are used by raising the load factor. The constraint of common origin and destination or at least identical routes and compatible time schedules of passengers limits the scope of car pools. (Another method of reducing pollution peaks, the staggering of work hours, would reduce the opportunities for car pools). The reasons why more car pools have not developed in the absence of government inducements are the same as those associated with driver insensitivity to increases in the cost of car usage. Assistance via provision of information, as well as regulation (such as restriction of free parking space to car pools), can be combined with differential tolls at rush hour (perhaps total exemption for car pools) to widen the cost difference between individual driving and car pools.

An alternative suggestion is subsidy to mass transit, whether such transportation is regarded as an infant industry and subsidies are provided until it achieves an economic load factor, a loyal clientele, and improved service, or continuing subsidies are provided. A case cannot be made for a mass-transit subsidy in order to reduce air pollution. Automobile riders have demonstrated an extremely low sensitivity to public-transit prices. An enormous subsidy would be required to reduce substantially the use of cars for commuting. Tolls and parking taxes on automobile commuting are far more effective than public-transit subsidies in reducing car usage, and more in line with the social costs of producing and consuming urban transportation.

Rationing

The pollution rights approach involves setting a global emission quota for an area and partitioning it among local polluters by selling "rations" to pollute. Regulation by rationing without resort to prices as an allocating mechanism has also been considered. Its virtue is that allocation need not be on the basis of ability to pay; it can give greater weight to the needs of the poor than can an auction system. Rationing is one means of encouraging car pools. In World War II, car-pool drivers were favored in gasoline allocations, although the purpose was conservation of gasoline rather than reduced air pollution.

There are high administrative costs in running a rationing system effectively and equitably. Previous experience suggests that it is only workable in emergencies, when the public is prepared to accept the necessity for such measures, and that the system tends to break down gradually with the development of black markets which reduce its equity and acceptability. In fact, a clear emergency is not here, and the shortage that makes rationing possible does not exist. Rationing of gasoline was once suggested by the EPA as a major factor in a plan to reduce pollutant emissions by cutting traffic. The enforcement of rationing over prolonged periods is costly and questionably effective. For short periods of very high air pollution, anticipated rationing could be circumvented by stockpiling. A simple moratorium on sales of gasoline during such short periods (while allowing for emergency needs, such as those of doctors) would avoid at least most of the administrative difficulties.

Related Policies: Energy Economy

Any policy reducing energy consumption also reduces air pollution associated with energy generation. Lowering thermostats, for instance, not only saves gas, oil, and coal, it reduces emissions of the pollutants associated with burning of these fuels. Requirements that new-model cars meet fuel economy standards reduce pollution in two ways: they reduce consumption of fuel, and they allow more complete combustion of the fuel consumed, which is one way of enhancing fuel economy. To the extent that fuel economy is the result of reduced weight of cars, energy (and pollution) is spared in the manufacturing process: less steel needs to be produced for each car. Associated policies to reduce the average weight of automobiles, such as the proposed tax on gas guzzlers and rebate on small cars, economize directly on gasoline consumption and indirectly on steel and other materials whose manufacture is associated with substantial air pollution. If the average American car weighed 2,500 pounds instead of 3,500, two million barrels of oil a day–12 percent of total consumption–would be saved.[17] The rough equivalent of car pools on scheduled airlines would economize on the number of flights and on fuel, thereby reducing this source of pollution, which is concentrated near the airports of large cities. Many scheduled flights are less than half full; cartels on major routes could raise the load factor significantly. Some economies in fuel consumption and reduction in pollution by trucks could be obtained by eliminating regulations that condemn nonregulated trucks to empty backhauls. (The initial negative effect on the average load factor of regulated trucks might have to be considered).

The energy crisis hit the United States in 1973, well after the Clean Air Act Amendments of 1970. Policies initially advocated to discourage use of cars because of their contribution to pollution, such as higher taxes on gasoline, were later advocated or implemented for the different but related objective of energy economy, independence, and conservation. Thus, insofar as the two goals are complementary, it is not easy to distinguish some policies for energy conservation from those for air pollution abatement. In choosing the best combination of policies, interdependencies both between goals and between policy instruments should be considered.

10 Summary

Our knowledge of air pollution and its effects is inadequate for the policies that are being formulated and implemented. Measurements for most pollutants in most areas do not extend far enough back in time. In most areas the number of instruments for measuring major pollutants is insufficient; the frequency as well as the spacing of observations is inadequate. Occasionally, if not often, the location of instruments is inappropriate for estimating pollution exposure. Measuring techniques, although improving, were inadequate during the period in which most health impact studies were conducted.

Even with detailed and accurate knowledge of air pollution, we have only a crude approximation of individual exposure. This depends on an individual's length of residence in various areas with diverse pollution experience; location of the individual's residence, work, and other major uses of time within an urban area, and method of traveling between them; the length of time the individual spends indoors; and the level and composition of indoor air pollution. The inadequacy of the evidence is particularly serious when health effects involve long lags or prolonged cumulative exposure. Reliable time series of air pollution just do not exist.

Measurements of the state of health may be as inadequate as measurements of exposure to air pollution. Mortality is the only clear-cut datum. Illnesses vary in degree of severity and in kind. We lack a consistent method for reporting degree of severity or even for classifying diseases by type. Hospital admissions, bed-days, doctors' visits, and days absent from work or school on account of illness are indirect estimates of poor health.

Both pollution and illness are conglomerate terms. The EPA provides an implicit weighting scheme for the several pollutants in terms of its air quality standards. Apart from serious reservations concerning the appropriateness of particular standards, air pollution almost always includes several pollutants that interact, so that standards set for each one as though it were the only pollutant around do not in fact calibrate its effect on health in the presence of other pollutants. Although some illnesses may facilitate the development of other illnesses, or aggravate them, the problem with measuring effects on health is that different pollutants may have qualitatively different health effects. Without some calculus to compare them, EPA or other standards for air pollutants may be equating eye irritation with lung cancer.

There have been hundreds of studies on the relation between air pollution and health. Most of them find a positive association between pollution levels and incidence of various illnesses or pathological symptoms. The reliability of the association is subject to the qualifications above on the very measurements which are associated. There are many other possible causes of observed differences in rates of disease or mortality, including age, sex, race, ethnic group, behavioral habits (including smoking and diet), and environmental differences between areas or time periods, such as climate, water supply, and domestic heating practices and facilities.

Although no one study holds constant all the variables that might explain observed associations between pollution and health, each variable is held constant by some studies. Thus, the studies in their entirety provide conclusive evidence that recent levels of air pollution in many of our cities are harmful to health. The evidence is strongest for sulfur dioxide and sulfates and particulates, and especially for the combination of the two. They are mainly of industrial and power-plant origin. The evidence is much weaker for NO_x. For CO, evidence of short-run effects is conclusive, but that of long-run effects is highly speculative. Most hydrocarbons can only be indicted as accessories to the photochemical production of oxidants. We remain unsure about the minimum threshold for damage to health. We are quite ignorant about the relation between level of pollution and frequency or severity of symptoms.

The poor quality of the evidence is not a basis for ignoring health effects. Although in some cases (nitrogen oxides, for instance) damage may have been greatly exaggerated, other health effects may have been understated or escaped notice. Our air pollution policies are better regarded as a hedge against uncertainty than as a premium based on precise calculations of costs and benefits.

For most sources of air pollution, large reductions in pollutant emission can be achieved at modest cost. All careful studies, notwithstanding their limitations, agree that the costs of achieving emission standards for sulfur oxides and particulates are less than the resulting benefits to health and life. This excess of health benefits over emission reduction costs strongly indicates that we should proceed to attain emission standards with present techniques. But motor vehicle emissions are a different story. The effects of NO_2, CO_2, and hydrocarbons on health are less well established than the effect of sulfur oxides and particulates. Present studies indicate that the cost of attaining motor vehicle emission standards exceeds the resulting health benefits. Under these circumstances, with the standards themselves in question, there is a case for relaxing and postponing emission standards while more effective and cheaper ways of reducing pollutant emissions are developed. In particular, the justification of the NO_x standards is highly questionable, the possibility of compliance on schedule is problematic, and the cost is very high. It should be possible to provide strong incentives for accelerated development and introduction of superior techniques without resort to the frontier device for wonderfully concentrating the mind: the imminent threat of hanging, which the 1975-76 deadline was meant to be.

In addition to the calculable costs of compliance with clean air laws, there are effects on productivity and growth whose future magnitude cannot be judged. These include some reduction in efficiency associated with use of pollution abatement equipment, and the possible substantial diversion of investment and research from capacity and productivity increases and new product development to pollution abatement. However, a judgment on pollution standards that counts all costs should also count all benefits. In addition to health, there are substantial benefits in reducing damage to materials and vegetation, as well as less tangible gains in aesthetics and quality of life. The net balance is unknown.

Regulation of air pollution has taken two forms: the setting of national air quality standards, which are directly pertinent to health, and the setting of national emission standards for motor vehicles and stationary sources of air pollution, which are only indirectly related to health. Recent controversies on permissible degradation of air quality in areas well within the standards and on control of land use to prevent excessive local concentration of emission sources affect location of economic activity and the regional distribution of economic growth.

The strict enforcement, on schedule, of ambient air quality standards would have stopped growth or induced decline in many of our larger cities. A trade-off policy allowing new polluting plants to locate averts this consequence, at the price of giving existing plants an incentive to minimize pollution abatement in the hope of benefiting from a trade-off. Nondegradation policies in areas with pollution levels well below national standards hamper the growth of economic activity, and cannot be justified in terms of avoiding health damage. Emission standards that are equally applicable to plants in clean and in dirty areas discourage the redistribution of industry so as to minimize health damage, and are not the least-cost approach to health improvement.

The regulatory approach selected does not allow each area, or each polluter, a choice. Other regulatory approaches, such as emission inspection and mandatory maintenance or a two-standard strategy for cars whereby the stricter emission standard is applicable only to polluted areas, can achieve the same air quality standards at much lower cost. An emission tax whose rate is set with desired air quality in view, or sale of a fixed amount of pollution rights, would allow each polluter to decide whether to bear the costs of polluting or not, and to choose the most cost-effective means of achieving desired pollution abatement. These strategies would be preferable for major polluters, whose performance can be monitored. Extensive use of an emission-tax approach imposes monitoring requirements that cannot be satisfied at a reasonable cost at this time. For large numbers of small polluters, whose emissions it may never be practicable to monitor, a fuel tax, requirements of use of low-pollutant fuels, or excise taxes on engines and other sources of emissions geared to their operational characteristics are the most practicable incentives for reducing emissions. Subsidies, in particular accelerated depreciation and tax credits for antipollution equipment, are also a second-best approach that biases responses toward the subsidized option.

The great increase in the price of oil and other fuels since 1973 and policies aimed at reducing dependence on imported oil have increased the costs of air pollution abatement, and also promise to raise the benefits. Increasing reliance on coal, our dirtiest fuel, accentuates the conflict between clean air and energy independence. Only those energy policies that encourage reduced use of energy are complementary with clean air policies. Large costs and long-range commitments are at stake. Any reasoned compromise between the goals of environmental quality, energy, and economic growth is likely to be better than an outcome that is the accidental sum of the many ad hoc confrontations already under way.

Notes

Chapter 1
Introduction

1. Harold J. Barnett, "Natural Environment and Growth," in *U.S. Economic Growth from 1976 to 1986: Prospects, Problems, and Patterns,* Congress of the United States, Joint Economic Committee, 95th Congress, 1st session (Washington: U.S. Government Printing Office, 1977), p. 10.

Chapter 2
Emission, Pollution, and Exposure

1. W.E. Wilson, Jr., D.F. Miller, and A. Levy, "The Effect of Fuel Composition on Atmospheric Aerosol Due to Auto Exhaust," *Journal of the Air Pollution Control Association,* vol. 23 (November 1973), p. 949.

2. R.E. Lee, Jr., "The Size of Suspended Particulate Matter in the Air," *Science,* vol. 178 (10 November 1972), pp. 567-575. See also David F.S. Natusch and John R. Wallace, "Urban Aerosol Toxicity: The Influence of Particle Size," *Science,* vol. 186 (22 November 1974), pp. 695-699.

3. National Academy of Sciences, *Implications of Environmental Regulation for Energy Production and Consumption,* vol. 6 (Washington: U.S. Government Printing Office, 1977), p. 100.

4. Thomas H. Maugh II, "Carbon Monoxide: Natural Sources Dwarf Man's Output," *Science,* vol. 177 (28 July 1972), pp. 338-339.

5. W.W.Kellogg, R.D. Cadle, E.R. Allen, A.L. Lazrus, and E.A. Martell, "The Sulfur Cycle," *Science,* vol. 175 (11 February 1972), pp. 587-596.

6. E. Robinson and R.C. Robbins, "Sources, Abundance, and Fate of Gaseous Atmospheric Pollutants," (mimeographed), Stanford Research Institute Project PR-6755, February 1968, pp. 74, 89.

7. R.A. Rasmussen, "Isoprene: Identified as a Forest-Type Emission to the Atmosphere," *Environmental Science and Technology,* vol. 4 (August 1970), pp. 667-671. Also, Thomas H. Maugh II, "Air Pollution: Where Do Hydrocarbons Come From?" *Science,* vol. 189 (25 July 1975), pp. 277-278.

8. *Man's Impact on the Global Environment,* Report of the Study of Critical Environmental Problems, Massachusetts Institute of Technology, Cambridge, Mass., July 1970.

9. D. Weiss, B. Whitten, and D. Leddy, "Lead Content of Human Hair (1871-1971)," *Science,* vol. 178 (6 October 1972), pp. 69-70.

10. Aidan Cockburn, Robin A. Barraco, Theodore A. Reyman, and William H. Peck, "Autopsy of an Egyptian Mummy," *Science,* vol. 187 (28 March 1975), pp. 1155-1160.

11. R.T. Barber, A. Viyajakumar, and F.A. Cross, "Mercury Concentrations in Recent and Ninety-Year-Old Benthopelagic Fish," *Science,* vol. 178 (10 November 1972) p. 637.

12. Barry Kramer, "Radioactive Particles in Cigaret [Sic] Smoke Tied to Lung Cancer," *Wall Street Journal,* November 19, 1975.

13. Gene Likens and F. Herbert Bormann, "Acid Rain: A Serious Regional Environmental Problem," *Science,* vol. 184 (11 June 1974), p. 1176.

14. R.S. Dhindsa and J.D. Bewley, "Photochemical Air Pollution: Transport From the New York City Area into Connecticut and Massachusetts," *Science,* vol. 191 (16 January 1976), pp. 179-182.

15. Likens and Bormann, "Acid Rain," pp. 1176-1179.

16. Robinson and Robbins, "Sources of Atmospheric Pollutants," p. 9. See also Kellogg and others, "The Sulfur Cycle," p. 593.

17. Bernard Weinstock and Hiromi Niki, "Carbon Monoxide Balance in Nature," *Science,* vol. 176 (21 April 1972), pp. 290-292.

18. F.B. Abeles, L.E. Craker, L.E. Forrence, and G.R. Leather, "Fate of Air Pollutants: Removal of Ethylene, Sulfur Dioxide, and Nitrogen Dioxide by Soil," *Science,* vol. 173 (3 September 1971), pp. 914-916.

19. T. Timothy Crocker, in U.S. Congress, Senate, Subcommittee on Air and Water Pollution of the Committee on Public Works, *Implementation of the Clean Air Act Amendments of 1970: Hearings, Part 3,* 92nd Congress, 2nd session, March 25, 1972, p. 979.

20. U.S. Congress, Senate, Subcommittee on Air and Water Pollution of the Committee on Public Works, *The Impact of Auto Emission Standards* (Report of the Staff), 93rd Congress, 1st session, October 1973, pp. 50-51, 105-119.

21. "EPA Postpones for Year Emission Ceiling for Nitrogen Oxides on 1976 Model Cars," *Wall Street Journal,* July 31, 1973.

22. George B. Morgan, Guntis Ozolins, and Elbert C. Tabor, "Air Pollution Surveillance Systems," *Science,* vol. 170 (16 October 1970), pp. 289-296. Also, Thomas H. Maugh II, "Air Pollution Instrumentation: A Trend Toward Physical Methods," *Science,* vol. 177 (25 August 1972), pp. 685-687, and "Air Pollution Instrumentation (II): The Glamour of Lasers," *Science,* vol. 177 (22 September 1972), pp. 1090-1093.

23. J.A. Hodgeson, W.A. McClenny, and P.L. Hanst, "Air Pollution Monitoring by Advanced Spectroscopic Techniques," *Science,* vol. 182 (19 October 1973), pp. 248-257. Also, L.B. Kreuzer, N.D. Kenyon, and C.K.N. Patel, "Air Pollution: Sensitive Detection of Ten Pollutant Gases by Carbon Monoxide and Carbon Dioxide Lasers," *Science,* vol. 177 (28 July 1972), pp. 347-349. Also, R.T. Menzies and M.S. Shumate, "Air Pollution: Remote Detection of Several

Pollutant Gases with a Laser Heterodyne Radiometer," *Science,* vol. 184 (3 May 1974), pp. 570-572.

24. U.S. Congress, Senate, Subcommittee on Air and Water Pollution of the Committee on Public Works, *Hearings on Nondegradation Policy of the Clean Air Act,* 93rd Congress, 1st session, July 24, 1973, p. 198.

25. Morgan et al., "Air Pollution Surveillance," pp. 291, 293. Also, Alan J. Hoffman, Thomas C. Curran, Thomas B. McMullen, William M. Cox, and William F. Hunt, Jr., "EPA's Role in Ambient Air Quality Monitoring," *Science,* vol. 190 (17 October 1975), pp. 243-248.

26. Russell Train, "The Quest for Environmental Indices," *Science,* vol. 178 (13 October 1972), p. 121.

27. U.S. Congress, *Impact of Auto Emission Standards,* p. 45.

28. Morgan et al., "Air Pollution Surveillance," p. 295.

29. George Hagevik, "Pollution Control," *Science,* vol. 181 (6 July 1973), p. 10.

30. Harold E. Neustadter and Steven M. Sidik, "On Evaluating Compliance with Air Pollution Levels 'Not to Be Exceeded More Than Once a Year'," *Journal of the Air Pollution Control Association,* vol. 24 (June 1974), pp. 559-563.

31. *Environmental Quality,* Third Annual Report of the Council on Environmental Quality (Washington: U.S. Government Printing Office, 1972), p. 11.

32. *Impact of Auto Emission Standards,* p. 51.

33. *Environmental Quality,* 1972, pp. 7-10, 34-44.

34. *Environmental Quality,* Fourth Annual Report of the Council on Environmental Quality (Washington: U.S. Government Printing Office, 1973), p. 275.

35. Merril Eisenbud, "Environmental Protection in the City of New York," *Science,* vol. 170 (13 November 1970), pp. 706-712.

36. *Environmental Quality,* Sixth Annual Report of the Council on Environmental Quality (Washington, U.S. Government Printing Office, 1975), p. 338.

37. *Environmental Quality,* Seventh Annual Report of the Council on Environmental Quality (Washington: U.S. Government Printing Office, 1976), p. 247.

38. Train, "Quest for Indices," p. 121.

39. U.S. Office of Science and Technology, *Cumulative Regulatory Effects on the Cost of Automotive Transportation,* Final Report of the Ad Hoc Committee, February 28, 1972, p. xx.

40. V.J. Schaefer, V.A. Mohnen, and V.R. Veirs, "Air Quality of American Homes," *Science,* vol. 178 (10 November 1972), pp. 173-175. Also, Lee, "Size of Particulate Matter," pp. 573-575.

41. C. Ray Thompson, Earl G. Hensel, and Gerrit Kats, "Outdoor-Indoor Levels of Six Air Pollutants," *Journal of the Air Pollution Control Association,* vol. 23 (October 1973), pp. 881-886.

42. Robert S. Foote, "Mercury Vapor Concentrations Inside Buildings," *Science,* vol. 177 (11 August 1972), pp. 513-514.

43. Robert J. Nauman, "Smoking and Air Pollution Standards," *Science,* vol. 182 (26 October 1973), pp. 334, 336.

Chapter 3
Effects of Air Pollution Exposure on Health

1. U.S. Congress, Senate, Subcommittee on Air and Water Pollution of the Committee on Public Works, *Hearings on Nondegradation Policy of the Clean Air Act,* 93rd Congress, 1st session, July 24, 1973, p. 198.

2. *Nondegradation Policy,* pp. 198-199.

3. For a review of conflicting studies, see Charles H. Connolly, *Air Pollution and Public Health,* 2nd edition (New York: The Dryden Press, 1973), pp. 38-44 and Appendix C. Also, Ralph Binder and Deborah Feinberg, *Estimates of Measurable Human Responses to Carbon Monoxide,* Institution for Social and Policy Studies Working Paper W4-7, (New Haven: Yale University Press, 1974). Also, Frank P. Grad et al., *The Automobile and the Regulation of Its Impact on the Environment* (Norman, Okla.: University of Oklahoma Press, 1975), pp. 42-46. Also, Lester B. Lave and Eugene P. Seskin, *Air Pollution and Human Health* (Baltimore: The Johns Hopkins University Press, 1977), pp. 305-311.

4. Connolly, *Air Pollution and Health,* pp. 44-46, 57.

5. Stephen M. Ayres, R. Evans, D. Licht, J. Griesbach, F. Reimold, E.F. Ferrand, and A. Criscitiello, "Health Effects of Exposure to High Concentrations of Automotive Emissions," *Archives of Environmental Health,* vol. 27 (September 1973), pp. 168-178.

6. Frank E. Speizer and Benjamin G. Ferris, Jr., "Exposure to Automobile Exhaust," *Archives of Environmental Health,* vol. 26 (June 1973), pp. 313-329.

7. Connolly, *Air Pollution and Health,* Appendix D.

8. Bertram D. Dinman, "Pathophysiologic Determinants of Community Air Quality Standards for Carbon Monoxide," *Journal of Occupational Medicine,* vol. 10 (September 1968), pp. 14-24.

9. Richard D. Stewart, E. Baretta, L.R. Platte, E.B. Stewart, and A.A. Rimm, "Carboxyhemoglobin Concentration in Blood from Donors in Chicago, Milwaukee, New York, and Los Angeles," *Science,* vol. 182 (28 December 1973), pp. 1362-1364.

10. Poul Astrup, Knud Kjelsen, and John Wanstrup, "Effects of Carbon Monoxide Exposure on the Arterial Walls," *Biological Effects of Carbon Monoxide,* vol. 174 of *Annals of the New York Academy of Sciences,* October 5, 1970, pp. 294-300.

11. Connolly, *Air Pollution and Health,* pp. 50-54, reviews a number of studies relevant to adaptation.

12. Laurence D. Fechter and Zoltan Annau, "Toxicity of Mild Prenatal Carbon Monoxide Exposure," *Science,* vol. 197 (12 August 1977), pp. 680-682.

13. Richard B. Engdahl, "A Critical Review of Regulations for the Control of Sulfur Oxide Emissions," *Journal of the Air Pollution Control Association,* vol. 23 (May 1973), pp. 364-375.

14. Wesley C.L. Hemeon, "A Critical Review of Regulations for the Control of Particulate Emissions," *Journal of the Air Pollution Control Association* vol. 23 (May 1973), pp. 376-387. Also, Theodore D. Sterling, Seymour B.V. Pollack, and J. Weinkam, "Measuring the Effect of Air Pollution on Urban Morbidity," *Archives of Environmental Health,* vol. 18 (April 1969), pp. 485-494.

15. Charles McJilton and Robert Frank, "Role of Relative Humidity in the Synergistic Effect of a Sulfur Dioxide-Aerosol Mixture on the Lung," *Science,* vol. 182 (2 November 1973), pp. 503-504.

16. Robert W. Buechley, Wilson B. Riggan, Victor Hasselbad, and John B. Van Bruggen, "SO_2 Levels and Perturbations in Mortality," *Archives of Environmental Health,* vol. 27 (September 1973), pp. 134-137.

17. In a paper presented at the June 1974 meeting of the Air Pollution Control Association, reported in "SO_2's Health Threat Questioned in Study," *New York State Environment,* July 1, 1974, pp. 1, 10.

18. Harry Heimann, "Trial Epidemiological Study of Episodic Air Pollution in Metropolitan Boston," *Archives of Environmental Health,* vol. 20 (February 1970), pp. 230-251. Also, Thomas A. Hodgson, "Short Term Effects of Air Pollution on Mortality in New York City," *Environmental Science and Technology* (July 1970), pp. 589-597.

19. G.C. McDonald and R.C. Schwing, "Instabilities of Regression Estimates Relating Air Pollution to Mortality," *Technometrics,* vol. 15 (August 1973), pp. 463-481.

20. Lave and Seskin, *Air Pollution and Human Health,* pp. 32-43, 120-125.

21. William H. Durham, "Air Pollution and Student Health," *Archives of Environmental Health,* vol. 28 (May 1974), pp. 241-254.

22. S. Ishikawa, D.H. Bowden, V. Fisher, and J.P. Wyatt, "The 'Emphysema Profile' of Two Midwestern Cities in North America," *Archives of Environmental Health,* vol. 18 (April 1969), pp. 660-666.

23. Herbert E. Stokinger and David L. Coffin, "Biologic Effects of Air Pollutants," in Arthur C. Stern (ed.), *Air Pollution,* 2nd edition (New York: Academic Press, 1968), pp. 445-546.

24. Carl M. Shy, Victor Hasselbad, Robert M. Burton, C.J. Nelson, and A.A. Coles, "Air Pollution Effects on Ventilatory Function of US Schoolchildren," *Archives of Environmental Health,* vol. 27 (September 1973, pp. 124-128.

25. U.S. Department of Health, Education and Welfare, *Air Quality Criteria for Sulfur Oxides,* National Air Pollution Control Publication no. AP-50, January 1969.

26. W.W. Kellogg, R.D. Cadle, E.R. Allen, A.L. Lazrus, and E.A. Martell, "The Sulfur Cycle," *Science,* vol. 175 (11 February 1972), p. 593. Also, *Cleaning Our Environment: The Chemical Basis for Action* (Washington: American Chemical Society, 1969), pp. 27-30.

27. Stokinger and Coffin, "Effects of Air Pollutants," pp. 472-474. Also, Steven B. Valand, Gene D. Acton, and Q.N. Myrvic, "Nitrogen Dioxide Inhibition of Viral Induced Resistance in Alveolar Macrophages," *Archives of Environmental Health,* vol. 10 (March 1965), pp. 303-309.

28. Hans L. Falk, "Chemical Definitions of Inhalation Hazards," *Inhalation Carcinogenesis,* Proceedings of the Biology Division, Oak Ridge National Laboratory, Gatlinburg, Tennessee, October 8-11, 1969 (Washington: U.S. Atomic Energy Commission, April 1970), Conf. 69-1001, pp. 13-26.

29. Carl M. Shy, John P. Creason, Martin E. Pearlman, Kathryn E. McClain, Ferris Benson, and Marion M. Young, "The Chattanooga School Children Study: Effects of Community Exposure to Nitrogen Dioxide," *Journal of the Air Pollution Control Association,* Part 1, vol. 20 (August 1970), pp. 539-545; Part 2, vol. 20 (September 1970), pp. 582-588.

30. Martin E. Pearlman, J.F. Finklea, John P. Creason, Carl M. Shy, Marion M. Young, and R.J.M. Horton, "Nitrogen Dioxide and Lower Respiratory Illness," *Pediatrics,* vol. 47 (February 1971), pp. 391-398.

31. Connolly, *Air Pollution and Health,* pp. 108-111; Grad et al., *The Automobile and the Environment,* pp. 46-47.

32. George W. Wright, "An Appraisal of Epidemiological Data Concerning the Effect of Oxidants, Nitrogen Dioxide, and Hydrocarbons upon Human Populations," paper presented at the 62nd Annual Meeting of the Air Pollution Control Association, New York, June 1969.

33. Lave and Seskin, *Air Pollution and Human Health,* pp. 145, 196-205.

34. Enviro Control, Inc., *A Study of Photochemical Pollutants and Their Health Effects,* summary reprinted in U.S. Congress, Senate, Subcommittee on Air and Water Pollution of the Committee of Public Works, *The Impact of Auto Emission Standards* (Report of the Staff), 93rd Congress, 1st session, October 1973, p. 61.

35. E.V. Sargent, R.T. Zagranski, and J.A.J. Stolwijk, *Estimates of Acute Respiratory Illness Due to Nitrogen Dioxides Above the Primary Standard,* Working Paper WC5-12, Institution for Social and Policy Studies, Yale University, 1975, p. 15.

36. William D. Wagner, Burris R. Duncan, Paul C. Wright, and Herbert E. Stokinger, "Environmental Study of Threshold Limit of NO_2," *Archives of Environmental Health,* vol. 10 (March 1965), pp. 455-456. Also, Arthur R.

Gregory and Charles H. Hine, "Neonatal Resistance to Lung Edema," *Proceedings of the Society for Experimental Biology and Medicine,* 1968, pp. 693-695.

37. J.C. McConnell, M.B. McElroy, and S.C. Wofsy, "Natural Sources of Atmospheric CO," *Nature,* vol. 233 (September 17, 1971), pp. 187-188. Some writers question whether polynuclear hydrocarbons cause cancer of the lung by themselves, or only in association with an irritant. For a discussion, see Connolly, *Air Pollution and Health,* pp. 86-88.

38. Connolly, *Air Pollution and Health,* pp. 123-135.

39. David L. Coffin, "Study of the Mechanisms of the Alteration of Susceptibility to Infection Conferred by Oxidant Air Pollutants," *Inhalation Carcinogenesis,* pp. 259-264.

40. Sterling et al., "Effect of Pollution on Morbidity," pp. 489-493.

41. W.S. Wayne, P.F. Wehrle, and R.E. Carrol, "Oxidant Air Pollution and Athletic Performance," *Journal of the American Medical Association,* vol. 199 (March 20, 1967), pp. 901-904. Also, Hans K. Ury, "Photochemical Air Pollution and Automobile Accidents in Los Angeles," *Archives of Environmental Health,* vol. 17 (September 1968), pp. 334-342.

42. C.E. Schoettlin and E. Landau, "Air Pollution and Asthmatic Attacks in the Los Angeles Area," *Public Health Reports,* 1961, pp. 545-548.

43. Enviro Control, Inc., *Photochemical Health Effects,* summary in *Impact of Auto Emission Standards,* p. 61.

44. U.S. Department of Health, Education and Welfare, *Air Quality Criteria for Photochemical Oxidants,* Publication no. AP-63 (Washington: National Air Pollution Control Administration, 1970).

45. Stephen M. Alpert and Trent R. Lewis, *Ozone Tolerance Studies: Unilateral Lung Exposure* (Washington: National Air Pollution Control Administration, 1970).

46. David F.S. Natush and John R. Wallace, "Urban Aerosol Toxicity: The Influence of Particle Size," *Science,* vol. 186 (22 November 1974), pp. 695-699.

47. Warren Winkelstein, Jr. and Michael L. Gay, "Arteriosclerotic Heart Disease and Cerebrovascular Disease: Further Observations on the Relationship of Suspended Particulate Air Pollution and Mortality in the Erie County Air Pollution Study," *Proceedings of the Institute of Environmental Studies,* 1970, pp. 441-447. Also, "Suspended Particulate Air Pollution: Relationship to Mortality from Cirrhosis of the Liver," *Archives of Environmental Health,* vol. 22 (January 1971), pp. 174-177.

48. Lave and Seskin, *Air Pollution and Human Health,* pp. 30-55, 120-159, 167-185.

49. M.C. Battigelli, "Sulfur Dioxide and Acute Effects of Air Pollution," *Journal of Occupational Medicine,* vol. 10 (September 1968), pp. 500-511.

50. Ralph J. Caprio, Harry L. Margulis, and Morris M. Joselow, "Lead Absorption in Children and Its Relation to Urban Traffic Densities," *Archives of Environmental Health,* vol. 28 (April 1974), pp. 195-197.

51. D. Weiss and B. Whitten, "Lead Content of Human Hair (1871-1971)," *Science,* vol. 178 (6 October 1972), pp. 69-70. Also, Martha McLaughlin, Adrian Linch, and Ronald D. Snee, "Longitudinal Studies of Lead Levels in a US Population," *Archives of Environmental Health,* vol. 27 (November 1973), pp. 305-311.

52. Michael Rabinowitz and George W. Wetherill, "Lead Metabolism in the Normal Human: Stable Isotope Studies," *Science,* vol. 182 (16 November 1973), pp. 725-727.

53. W.J. Nicholson and A. Rohl, "Asbestos Air Pollution in New York City," paper presented at the 2nd International Clean Air Congress of the International Union of Air Pollution Prevention Associations, Washington, D.C., December 1970. Also, Arthur N. Rohl, Arthur M. Langer, and Irving J. Selikoff, "Environmental Asbestos Pollution Related to Use of Quarried Serpentine Rock," *Science,* vol. 196 (17 June 1977), pp. 1319-1322.

54. Frank R. Dutra and John D. Carney, "Asbestosis and Pulmonary Carcinoma," *Archives of Environmental Health,* vol. 10 (March 1965), pp. 416-423.

55. E.C. Hammond and D. Horn, "Smoking and Death Rates—Report on Forty-four Months of Follow-up of 187,783 Men: II. Death Rates by Cause," *Journal of the American Medical Association,* vol. 166 (March 15, 1958), pp. 1294-1308. Also, W. Haenszel, D.B. Loveland, and M.G. Sirken, "Lung Cancer Mortality as Related to Residence and Smoking Histories: I. White Males," *Journal of the National Cancer Institute,* vol. 28 (April 1962), pp. 947-1001. Also, M.L. Levin, W. Haenszel, B.E. Carroll, P.R. Gerhardt, V.H. Handy, and S.C. Ingraham, "Cancer Incidence in Urban and Rural Areas of New York State," *Journal of the National Cancer Institute,* vol. 24 (June 1960), pp. 1243-57.

56. Haenszel et al., "Lung Cancer Mortality," pp. 963-973. Also, P. Buell, J.E. Dunn, and L. Breslow, "Cancer of the Lung and Los Angeles-Type Air Pollution," *Cancer,* vol. 20 (December 1967), pp. 2139-2147.

57. C.A. Mills, "Urban Air Pollution and Respiratory Diseases," *American Journal of Hygiene,* vol. 37 (March 1943), pp. 131-141. Also, "Motor Exhaust Gases and Lung Cancer in Cincinnati," *American Journal of Medical Science,* vol. 239 (March 1960), pp. 316-319.

58. C.A. Mills, "Air Pollution and Community Health," *American Journal of Medical Science,* vol. 224 (October 1952), pp. 403-407.

59. P. Stocks and J.M. Campbell, "Lung Cancer Death Rates Among Non-smokers and Pipe and Cigarette Smokers, An Evaluation in Relation to Air Pollution by Benzopyrene and Other Substances," *British Medical Journal,* vol. 2 (October 15, 1955), pp. 923-929. Also C. Daly, "Air Pollution and Causes of

Death," *British Journal of Preventive and Social Medicine,* vol. 13 (January 1959), pp. 14-27. Also, M.P. Curwen and E.L. Kennaway, "The Incidence of Cancer of the Lung and Larynx in Urban and Rural Districts," *British Journal of Cancer,* vol. 8 (June 1954), pp. 181-198. Also, L. Kreyberg, "Occurrence and Aetiology of Lung Cancer in Norway in the Light of Pathological Anatomy," *British Journal of Preventive and Social Medicine,* vol. 10 (October 1956), pp. 145-158.

60. G. Dean, "Lung Cancer in Australia," *Medical Journal of Australia,* vol. 1 (June 30, 1962), pp. 1003-1006. Also, "Lung Cancer Among White South Africans: Report on a Further Study," *British Medical Journal,* vol. 2 (December 16, 1961), pp. 1599-1605.

61. D.F. Eastcott, "The Epidemiology of Lung Cancer in New Zealand," *Lancet,* vol. 1 (January 7, 1956), pp. 37-39.

62. Lave and Seskin, *Air Pollution and Human Health,* p. 296.

63. W. Winkelstein, Jr. and S. Kantor, "Stomach Cancer: Positive Association with Suspended Particulate Air Pollution," *Archives of Environmental Health,* vol. 18 (April 1969), pp. 544-547. Also, "Prostatic Cancer: Relationship to Suspended Particulate Air Pollution," *American Journal of Public Health,* vol. 59 (July 1969), pp. 1134-1138. Also, Curwen and Kennaway, "Cancer of the Lung," pp. 181-198, and Mills, "Pollution and Respiratory Diseases," pp. 137-139. See also T.F. Mancuso, E.M. McFarlane, and J.D. Porterfield, "Distribution of Cancer Mortality in Ohio," *American Journal of Public Health,* vol. 45 (January 1955), pp. 58-70. Also, M.J. Gardner, M.D. Crawford, and J.N. Morris, "Patterns of Mortality in Middle and Early Old Age in the County Boroughs of England and Wales," *British Journal of Preventive and Social Medicine,* vol. 23 (August 1969), pp. 133-140. Also, R.M. Hagstrom, H.A. Sprague, and E. Landau, "The Nashville Air Pollution Study VII: Mortality from Cancer in Relation to Air Pollution," *Archives of Environmental Health,* vol. 15 (August 1967), pp. 237-247.

64. Bertram W. Carnow and Paul Meier, "Air Pollution and Pulmonary Cancer," *Archives of Environmental Health,* vol. 27 (September 1973), pp. 207-218.

65. Herman R. Menck, John T. Casagrande, and Brian E. Henderson, "Industrial Air Pollution: Possible Effect on Lung Cancer," *Science,* vol. 183 (18 January 1974), pp. 210-212.

66. Connolly, *Air Pollution and Health,* pp. 237-238.

67. *Nondegradation Policy,* p. 213.

68. *Nondegradation Policy,* pp. 217-218.

69. Hugues J.-P. Ryser, "Chemical Carcinogens," *New England Journal of Medicine,* vol. 285 (September 23, 1971), pp. 721-734.

70. W. Winkelstein, Jr. and S. Kantor, "Respiratory Symptoms and Air Pollution in an Urban Population of Northeastern United States," *Archives of Environmental Health,* vol. 18 (May 1969), pp. 760-767.

71. L.D. Zeidberg, R.J.M. Horton, and E. Landau, "The Nashville Air Pollution Study V: Mortality from Diseases of the Respiratory System in Relation to Air Pollution," *Archives of Environmental Health,* vol. 15 (August 1967), pp. 214-224.

72. M.H. Lepper, N. Shioura, B. Carnow, S. Andelman, and L. Lehrer, "Respiratory Disease in an Urban Environment," *Industrial Medicine and Surgery,* vol. 38 (April 1969), pp. 126-131.

73. Shy et al., "The Chattanooga Study, Part 2," pp. 582-588.

74. Pearlman et al., "Respiratory Illness," pp. 391-398.

75. C.E. Schoettlin, "The Health Effect of Air Pollution on Elderly Males," *American Review of Respiratory Disease,* vol. 86 (December 1962), pp. 878-897.

76. H.L. Motley, R.A. Smart, and C.I. Leftwich, "Effect of Polluted Los Angeles Air (Smog) on Lung Volume Measurements," *Journal of the American Medical Association,* vol. 171 (November 14, 1959), pp. 1469-1477.

77. F.E. Speizer and B.C. Ferris, Jr., "The Prevalence of Chronic Non-Specific Respiratory Disease in Road Tunnel Employees," *American Review of Respiratory Disease,* vol. 88 (August 1963), pp. 205-212.

78. J.W.B. Douglas and R.E. Waller, "Air Pollution and Respiratory Infection in Children," *British Journal of Preventive and Social Medicine,* vol. 20 (January 1966), pp. 1-8. Also, J.R.T. Colley and D.D. Reid, "Urban and Social Origins of Childhood Bronchitis in England and Wales," *British Medical Journal,* vol. 2 (April 1970), pp. 213-217.

79. W.W. Holland, D.D. Reid, R. Seltser, and R.W. Stone, "Respiratory Disease in England and the United States," *Archives of Environmental Health,* vol. 10 (February 1965), pp. 338-343. Also, J.E. Lunn, J. Knowelden, and J.W. Roe, "Patterns of Respiratory Illness in Sheffield Junior School Children: A Follow-Up Study," *British Journal of Preventive and Social Medicine,* vol. 24 (November 1970), pp. 223-228.

80. R.L. Petrilli, G. Agnese, and S. Kanitz, "Epidemiologic Studies of Air Pollution Effects in Genoa, Italy," *Archives of Environmental Health,* vol. 12 (June 1966), pp. 733-740.

81. T. Toyama, "Air Pollution and Health Effects in Japan," *Archives of Environmental Health,* vol. 8 (January 1964), pp. 153-173. Also, Y. Oshima, T. Ishizaki, T. Miyamoto, T. Shimizu, T. Shida, and J. Kabe, "Air Pollution and Respiratory Diseases in the Tokyo-Yokohama Area," *American Review of Respiratory Disease,* vol. 90 (October 1964), pp. 572-581.

82. I.O. Skalpe, "Long Term Effects of Sulfur Dioxide Exposure in Pulp Mills," *British Journal of Industrial Medicine,* vol. 21 (January 1964), p. 69.

83. J.J. Collins, J.S. Kasap, and W.W. Holland, "Environmental Factors in Child Mortality in England and Wales," *American Journal of Epidemiology,* vol. 93 (January 1971), pp. 10-22.

84. P. Stocks, "Cancer and Bronchitis Mortality in Relation to Atmo-

spheric Deposit and Smoke," *British Medical Journal,* vol. 1 (January 10, 1959), pp. 74-79, and "On the Relations Between Atmospheric Pollution in Urban and Rural Localities and Mortality from Cancer, Bronchitis and Pneumonia, with Particular Reference to 3:4 Benzopyrene, Beryllium, Molybdenum, Vanadium and Arsenic," *British Journal of Cancer,* vol. 14 (September 1960), pp. 397-418. Also, D.J.B. Ashley, "The Distribution of Lung Cancer and Bronchitis in England and Wales," *British Journal of Cancer,* vol. 21 (June 1967), pp. 243-259, and "Environmental Factors in the Aetiology of Gastric Cancer," *British Journal of Preventive and Social Medicine,* vol. 23 (November 1969), pp. 258-262. See also E. Gorham, "Pneumonia and Atmospheric Sulphate Deposit," *Lancet,* vol. 2 (September 5, 1959), pp. 287-288.

85. S.F. Buck and A.J. Wicken, "Models for Use in Investigating the Risk of Mortality from Lung Cancer and Bronchitis," *Applied Statistics Series C,* no. 3 (1967), pp. 185-201.

86. D. Hewitt, "Mortality in the London Boroughs, 1950-52, with Special Reference to Respiratory Disease," *British Journal of Preventive and Social Medicine,* vol. 10 (April 1956), pp. 45-57.

87. Lave and Seskin, *Air Pollution and Human Health,* pp. 275-277, 282.

88. A.S. Fairbairn and D.D. Reid, "Air Pollution and Other Local Factors in Respiratory Disease," *British Journal of Preventive and Social Medicine,* vol. 12 (April 1958), pp. 94-103. Also, D.D. Reid, "Environmental Factors in Respiratory Disease," *Lancet,* vol. 1 (June 21, 1958), pp. 1289-1294. Also, W.W. Holland and D.D. Reid, "The Urban Factor in Chronic Bronchitis," *Lancet,* vol. 1 (February 27, 1965), pp. 445-448.

89. K. Yoshida, H. Oshima, and M. Imai, "Air Pollution and Asthma in Yokkaichi," *Archives of Environmental Health,* vol. 13 (December 1966), pp. 763-768. Also, K. Yoshida, Y. Takatsuka, M. Kitabatake, H. Oshima, and M. Imai, "Air Pollution and its Health Effects in the Yokkaichi Area: Review on Yokkaichi Asthma," *Mie Medical Journal,* vol. 18 (January 1969), pp. 195-209.

90. R. Cederlof, "Urban Factor and Prevalence of Respiratory Symptoms and Angina Pectoris," *Archives of Environmental Health,* vol. 13 (December 1966), pp. 743-748.

91. P. Stocks, "Lung Cancer and Bronchitis in Relation to Cigarette Smoking and Fuel Consumption in Twenty Countries," *British Journal of Preventive and Social Medicine,* vol. 21 (October 1967), pp. 181-185.

92. Mills, "Pollution and Respiratory Diseases," pp. 133-138.

93. Mills, "Pollution and Community Health," pp. 404-406.

94. Stocks, "Pollution and Mortality," pp. 413-417. Also, Fairbairn and Reid, "Pollution and Respiratory Disease," pp. 99-102, and Gorham, "Pneumonia and Sulphate Deposit," pp. 287-288.

95. Daly, "Pollution and Causes of Death," pp. 19, 22. Also, Hewitt, "Mortality in London," p. 55.

96. Ishikawa et al., "Emphysema Profile," pp. 662-665.

97. H.A. Sultz, J.G. Feldman, E.R. Schlesinger, and W.E. Mosher, "An Effect of Continued Exposure to Air Pollution on the Incidence of Chronic Childhood Allergic Disease" (Paper presented at the 97th Annual Meeting of the American Public Health Association, Philadelphia, Pa., November 11, 1969).

98. Lave and Seskin, *Air Pollution and Human Health,* pp. 69, 74.

99. N.E. Manos and G.F. Fisher, "An Index of Air Pollution and Its Relation to Health," *Journal of the Air Pollution Control Association,* vol. 9 (January 1959), pp. 5-11.

100. L.D. Zeidberg, R.A. Prindle, and E. Landau, "The Nashville Air Pollution Study III: Morbidity in Relation to Air Pollution," *American Journal of Public Health,* vol. 54 (January 1964), pp. 85-97.

101. Zeidberg et al., "Nashville V: Respiratory Mortality," pp. 214-224.

102. Hodgson, "Air Pollution Mortality in New York," pp. 589-597.

103. H.T. Sauer, G.H. Payne, C.R. Council, and J.C. Terrell, "Cardio-Vascular Disease Mortality Patterns in Georgia and North Carolina," *Public Health Reports,* vol. 81 (May 1966), pp. 455-465. Also, P.E. Enterline, A.E. Rikli, H.T. Sauer, and M. Hyman, "Death Rates for Coronary Heart Disease in Metropolitan and Other Areas," *Public Health Reports,* vol. 75 (August 1960), pp. 759-766. Also, G.D. Friedman, "Cigarette Smoking and Geographic Variation in Coronary Hearth Disease Mortality in the U.S.," *Journal of Chronic Diseases,* vol. 20 (October 1967), pp. 769-779.

104. Gardner et al., "Patterns of Mortality in England and Wales," pp. 133-140.

105. Cederlof, "Urban Factor and Angina," pp. 745-747.

106. U.S. Secretary of Health, Education and Welfare, *National Emission Standards Study,* Report to the U.S. Congress (Senate Document No. 91-63, March 1970), pp. 58-59.

107. Lave and Seskin, *Air Pollution and Human Health,* p. 218.

108. Lave and Seskin, *Air Pollution and Human Health,* pp. 44-48, 127-132.

109. Connolly, *Air Pollution and Health,* Appendix I.

110. C.A. Cohen, A.R. Hudson, J.L. Clausen, and J.H. Knelson, "Respiratory Symptoms, Spirometry, and Oxidant Air Pollution in Nonsmoking Adults," *American Review of Respiratory Disease,* vol. 105 (February 1972), pp. 251-261.

111. A.E. Martin and W.H. Bradley, "Mortality, Fog and Atmospheric Pollution," *Monthly Bulletin,* Ministry of Health (Great Britain), vol. 19 (1960), p. 56, cited in Lave and Seskin, *Air Pollution and Health,* p. 211.

112. Hodgson, "Air Pollution Mortality in New York," pp. 594-96. Also, J. McCarroll and W. Bradley, "Excess Mortality as an Indicator of Health Effects of Air Pollution," *American Journal of Public Health,* vol. 56 (November 1966), pp. 1933-1942.

113. Lave and Seskin, *Air Pollution and Human Health,* p. 211.

Chapter 4
The Regulation of Air Pollution

1. U.S. Congress, Senate, Committee on Public Works, *National Air Quality Standards Act of 1970* (Report no. 91-1196), 91st Congress, 2nd session, September 17, 1970, p. 20.

2. *Environmental Quality,* Seventh Annual Report of the Council on Environmental Quality (Washington: U.S. Government Printing Office, 1976), p. 4.

3. National Academy of Sciences, Committee on Motor Vehicle Emissions, *Semiannual Report in Accordance with Section 202(c) of the Clean Air Amendments of 1970,* January 1, 1972, p. 6.

4. *Federal Register,* vol. 36, no. 247 (December 23, 1971), pp. 24876–24895, and vol. 37, no. 55 (March 21, 1972), pp. 5767–5771.

5. "EPA to Regulate Seven New Stationary Sources of Air Pollution, But Not Rendering Plants," *Wall Street Journal,* June 6, 1973.

6. Burt Schorr, "EPA Plans Revised Sulphur Dioxide Curbs that Could Help Some Utilities, Smelters," *Wall Street Journal,* June 28, 1973.

7. Wayne Green, "Justices Affirm the EPA Must Bar States from Allowing Deterioration of Clean Air," *Wall Street Journal,* June 12, 1973, p. 3.

8. Burt Schorr, "EPA Pushing to Control 'Complex' Sources of Air Pollution Despite Some Misgivings," *Wall Street Journal,* April 16, 1973, and "Clean Air Rules on City Projects Adopted by EPA," *Wall Street Journal,* June 12, 1973.

9. William H. Jones, "U.S. Accelerates Fuel Economy Rules for Cars," *The Washington Post,* June 27, 1977.

10. Richard O. Zerbe and Kevin Croke, *Urban Transportation for the Environment* (Cambridge, Mass.: Ballinger, 1975), p. 11.

11. *Environmental Quality,* 1976, p. 237.

12. *Environmental Quality,* 1976, pp. 216, 225–226, 231, 238–239.

13. U.S. Congress, Senate, Subcommittee on Air and Water Pollution of the Committee on Public Works, *1970 Hearings–Air Pollution, Part 4,* 91st Congress, 2nd session, p. 1288.

14. U.S. Secretary of Health, Education and Welfare, *National Emission Standards Study,* Report to the U.S. Congress (Senate Document No. 91-63, March 1970), p. 59.

15. T. Timothy Crocker, in U.S. Congress, Senate, Subcommittee on Air and Water Pollution of the Committee on Public Works, *Implementation of the Clean Air Act Amendments of 1970: Hearings, Part 3,* 92nd Congress, 2d session, March 25, 1972, p. 978.

16. U.S. Office of Science and Technology, *Cumulative Regulatory Effects on the Cost of Automotive Transportation* (RECAT), Final Report of the Ad Hoc Committee, February 28, 1972, pp. 6–7.

17. D.S. Barth, "Federal Motor Vehicle Emission Goals for Carbon Monoxide, Hydrocarbons, and Nitrogen Oxides Based on Desired Air Quality Levels," *Journal of the Air Pollution Control Association,* vol. 20 (August 1970), pp. 519-523. See also U.S. Congress, Senate, Subcommittee on Air and Water Pollution of the Committee on Public Works, *The Impact of Auto Emission on Standards* (Report of the staff), 93rd Congress, 1st session, October 1973, Introduction, Appendix C, pp. 17-27.

18. See, for example, the statement of Prof. Timothy Crocker and "Summary of Discussions at the National Motor Vehicle Air Pollution Conference," Western White House, San Clemente, California, January 13-14, 1972, in *Implementation of the Clean Air Act Amendments of 1970: Hearings Part 3,* pp. 977-979, 1327-1330. See also RECAT Report, p. 5.

19. United States Environmental Protection Agency, Office of Research and Monitoring, *EXPRO '73–A Listing of Extramural Projects to be Funded in Fiscal Year 1973,* Washington, D.C., October 1972, pp. 13-18.

20. *Cleaning Our Environment: The Chemical Basis for Action* (Washington: American Chemical Society, 1969), pp. 76-77.

21. RECAT Report, p. x.

Chapter 5
Benefits of Pollution Abatement

1. George B. Morgan, Guntis Ozolins, and Elbert C. Tabor, "Air Pollution Surveillance Systems," *Science,* vol. 170 (16 October 1970), p. 295.

2. Thomas C. Schelling, "The Life You Save May Be Your Own," in Samuel B. Chase (ed.), *Problems in Public Expenditure Analysis* (Washington: The Brookings Institution, 1968), pp. 127-162.

3. Ronald G. Ridker, *Economic Costs of Air Pollution* (New York: Praeger, 1967).

4. Lester B. Lave and Eugene B. Seskin, "Air Pollution and Human Health," *Science,* vol. 169 (21 August 1970), pp. 723-733.

5. Larry B. Barrett and Thomas E. Waddell, *The Cost of Air Pollution Damages: A Status Report* (Environmental Protection Agency: Research Triangle Park, N.C.: 1973). For a summary of other estimates see *Environmental Quality,* Sixth Annual Report of the Council on Environmental Quality (Washington, U.S. Government Printing Office, 1975), pp. 516-518.

6. U.S. Office of Science and Technology, *Cumulative Regulatory Effects on the Cost of Automotive Transportation* (RECAT), Final Report of the Ad Hoc Committee, February 28, 1972, p. 27. There are several other estimates, most of them intermediate between the two cited, although one EPA estimate (see RECAT, p. xix) is lower. See *Environmental Quality,* 1975, p. 517 for references.

7. Lester B. Lave and Eugene P. Seskin, *Air Pollution and Human Health* (Baltimore: The Johns Hopkins University Press, 1977) pp. 222-226.

8. Merril Eisenbud, "Environmental Protection in the City of New York," *Science,* vol. 170 (13 November 1970), p. 708.

9. Ronald G. Ridker and John A. Henning, "The Determinants of Residential Property Values with Special Reference to Air Pollution," *Review of Economics and Statistics,* vol. 44 (1967), pp. 246-257.

10. Robert J. Anderson and Thomas D. Crocker, "Air Pollution and Residential Property Values," *Urban Studies,* vol. 8, no. 3 (October 1971), pp. 171-180.

11. C. John Langley, Jr., "Adverse Impacts of the Washington Beltway on Residential Property Values," *Land Economics,* vol. 52 (February 1976), pp. 54-64.

12. David Harrison, Jr. and Daniel L. Rubinfeld, "Hedonic Housing Prices and the Demand for Clean Air," *Journal of Environmental Economics and Management,* vol. 5, no. 1 (March 1978), pp. 95-98.

13. Ridker, *Economic Costs of Air Pollution,* pp. 57-89.

14. V. Kerry Smith and Timothy A. Deyak, "Measuring the Impact of Air Pollution on Property Values," *Journal of Regional Science,* vol. 15 (December 1975), pp. 277-288.

15. A. Myrick Freeman III, "Air Pollution and Property Values: A Methodological Comment," *Review of Economics and Statistics,* vol. 53, (November 1971), pp. 415-416. Also, Hugh O. Nourse, "The Effect of Air Pollution on House Values," *Land Economics,* vol. 43 (May 1967), pp. 181-189.

16. Thomas E. Waddell, *The Economic Damages of Air Pollution,* EPA-600/5-74-012 (Washington, D.C.: U.S. Government Printing Office, 1974), reproduced in *Environmental Quality,* 1975, p. 506.

17. Jean L. Marx, "Air Pollution: Effects on Plants," *Science,* vol. 187 (28 February 1975), pp. 731-733.

18. *Environmental Quality,* 1975, p. 507.

19. Walter E. Westman, "How Much are Nature's Services Worth?" *Science,* vol. 197 (2 September 1977), pp. 960-964.

Chapter 6
Costs of Pollution Abatement

1. U.S. Congress, Senate, *The Economics of Clean Air,* Annual Report of the Administrator of the Environmental Protection Agency to the Congress of the United States, March 16, 1971, p. 2:7.

2. *Environmental Quality,* Fourth Annual Report of the Council on Environmental Quality (Washington: U.S. Government Printing Office, 1973) p. 83; also, Sixth Annual Report, 1975, p. 526.

3. *Environmental Quality,* Seventh Annual Report of the Council on Environmental Quality, 1976, p. 354.

4. "Mechanics Hurt Fight Against Car Pollution as Drivers Seek Pep," *Wall Street Journal,* June 22, 1972. Also, "EPA Says VW Unit, 12 Dealers Tampered with Recall Repairs," *Wall Street Journal,* March 14, 1978.

5. Burt Schorr, "States' Tests Indicate Car Pollution Controls Don't Work Very Well," *Wall Street Journal,* December 14, 1972.

6. *Wall Street Journal,* December 14, 1972, quotes John Goodman of the California Consumer Affairs Department, the licensing agency for 9,600 Class A garages, as "warning" such repair shops that "if these stations aren't doing a proper job, we will have to take some licenses away."

7. *Environmental Quality,* 1973, pp. 93, 99; also 1976, p. 145.

8. U.S. Congress, *The Cost of Clean Air,* Annual Report of the Administrator of the Environmental Protection Agency to the Congress of the United States (Document No. 93-122), 93d Congress, 2d session, October 1974, p. 1:3.

9. U.S. Office of Science and Technology, *Cumulative Regulatory Effects on the Cost of Automotive Transportation* (RECAT), Final Report of the Ad Hoc Committee, February 28, 1972, pp. 24, 74.

10. Lee A. Iacocca, "Emission Standards: A Five-Year Pause," *Washington Post,* May 26, 1975.

11. *The Economics of Clean Air,* 1972, pp. 1:8-1:9, 5:5-5-15.

12. *Environmental Quality,* 1975, p. 527.

13. "Smokestack Scrubbers, Still Opposed by Some, are Proving Feasible," *Wall Street Journal,* June 14, 1977.

14. *Environmental Quality,* 1976, p. 160.

15. Robert E. Kohn points this out in "Price Elasticities of Demand and Air Pollution Control," *Review of Economics and Statistics,* vol. 54 (November 1972) pp. 392-400.

16. "EPA States Rule to Curb Lead Pollution of the Air–Smelters Will be Hardest Hit," *Wall Street Journal,* December 13, 1977.

17. *Environmental Quality,* 1976, p.160.

18. Ronald Ridker, "Population and Pollution in the United States," *Science,* vol. 176 (9 June 1972), pp. 1085-1090.

19. *Environmental Quality,* 1976, pp. 148-150.

20. Ralph d'Arge provides an illustration of the possible effect of environmental controls on the domestic income of five countries. It is minimal on the United States, small on West Germany, but possibly quite large on Japan, France, and the United Kingdom. Major determinants of the effect include the importance of foreign exports to domestic output (least for the United States) and the type of product exported. Exports of products whose competitive position is affected by environmental standards (for instance, steel) are less important to the United States than to most other advanced nations. See Ralph C. d'Arge, "Essay on Economic Growth and Environmental Quality," *The Swedish Journal of Economics,* vol. 43, no. 1 (March 1973), pp. 37-40.

21. G. Christian Hill, "Power Play," *Wall Street Journal,* September 12, 1978.

22. Walter Dolde, Dennis Epple, Milton Harris, Lester Lave, and Samuel Leinhardt, "Dynamic Aspects of Air Quality Control Costs," *Journal of Environmental Economics and Management,* vol. 4, no. 4 (December 1977), pp. 313–334.

23. National Academy of Sciences, Committee on Motor Vehicle Emissions, *Semiannual Report in Accordance with Section 202(c) of the Clean Air Amendments of 1970,* January 1, 1972, p. 49.

24. National Research Council, Division of Engineering, *Report by the Committee on Motor Vehicle Emissions,* February 12, 1973, pp. 2–3.

25. RECAT Committee Report, p. xi.

26. Frank P. Grad et al., *The Automobile and the Regulation of Its Impact on the Environment,* (Norman, Okla.: University of Oklahoma Press, 1975), p. 319.

27. "Union Oil Calls GM Bid for Reduced Volatility of Gasoline Too Costly," *Wall Street Journal,* October 9, 1972.

28. RECAT Committee Report, p. xxiii.

29. William D. Balgord, "Fine Particles Produced from Automotive Emissions Control Catalysts," *Science,* vol. 180 (15 June 1973), pp. 1168–1169.

30. Deborah Shapley, "Auto Pollution: EPA Worrying that Catalyst May Backfire," *Science,* vol. 182 (26 October 1973), pp. 368-371.

31. S.H. Cadle, D.P. Chock, J.M. Heuss, and P.R. Manson, "Results of the General Motors Sulfate Dispersion Experiment," General Motors Research Laboratories GMR-2107. Also, Thomas H. Maugh II, "Sulfuric Acid From Cars: A Problem that Never Materialized," *Science,* vol. 198 (21 October 1977), pp. 280-284. Also, National Bureau of Economic Research, *National Bureau Report,* New York, February 1975, p. 2.

32. Douglas Martin, "EPA Initially to Propose Tough Controls on Sulphur From New Coal-Fired Plants," *Wall Street Journal,* August 24, 1978. See also Edwin McDowell, "The Big Battle Over Scrubbers," *Wall Street Journal,* February 7, 1975.

33. National Bureau of Economic Research, *National Bureau Report,* New York, February 1975, pp. 2–3.

Chapter 7
Geographical Distribution of the Costs of Clean Air

1. Douglas Martin, "Curb on Construction Where Air is Dirty Rankles Businessmen," *Wall Street Journal,* May 4, 1977.

2. *Environmental Quality,* Seventh Annual Report of the Council on Environmental Quality (Washington: U.S. Government Printing Office, 1976), p. 3.

3. "Dirty Air Leads to Strange Industry 'Tradeoffs,'" *Washington Post,* May 15, 1977.

4. Scott E. Atkinson and Donald H. Lewis, "A Cost Effectiveness Analysis of Alternative Air Quality Control Strategies," *Journal of Environmental Economics and Management,* vol. 1, no. 3 (November 1974), pp. 237-250.

Chapter 8
Energy-Environment Trade-Offs

1. "Energy Enigma: Increasing Use of Coal Faces a Host of Problems," *Wall Street Journal,* June 9, 1977.

2. Walter S. Mossberg, "Solar Power Seen Meeting 20 Percent of Needs by 2000; Carter May Seek Outlay Boost," *Wall Street Journal,* August 22, 1978.

3. David E. Lilienthal, "Lost Megawatts Flow over Nation's Myriad Spillways," *Smithsonian,* September 1977, pp. 82-89.

4. *Environmental Quality,* Seventh Annual Report of the Council on Environmental Quality (Washington: U.S. Government Printing Office, 1976), pp. 205-207. For additional details and slightly different comparative efficiencies, see *Environmental Quality,* Sixth Annual Report, 1975, pp. 433-435.

5. A contrary view has been widely held. See Sherwood B. Idso and Anthony J. Brazel, "Planetary Radiation Balance as a Function of Atmospheric Dust: Climatological Consequences," *Science,* vol. 198 (18 November 1977), pp. 731-733.

6. U. Siegenthaler and H. Oeschger, "Predicting Future Atmospheric Carbon Dioxide Levels," *Science,* vol. 199 (27 January 1978), pp. 388-395.

7. Jerzy Neyman, "Public Health Hazards from Electricity-Producing Plants," *Science,* vol. 195 (25 February 1977), pp. 754-758. Also, Christoph Hohenemser, Roger Kasperson, and Robert Kates, "The Distrust of Nuclear Power," *Science,* vol. 196 (1 April 1977), pp. 25-34.

8. Todd R. La Porte, "Nuclear Waste: Increasing Scale and Sociopolitical Impacts," *Science,* vol. 201 (7 July 1978), pp. 22-28.

9. J. Karkheck, J. Powell, and E. Beardsworth, "Prospects for District Heating in the United States," *Science,* vol. 195 (11 March 1977), pp. 948-955.

10. Ronald K. White and Richard L. Goen, "Cogeneration in West Germany," *Science,* vol. 197 (12 August 1977), p. 618.

11. Lee Schipper and Allan J. Lichtenberg, "Efficient Energy Use and Well-Being: The Swedish Example," *Science,* vol. 194 (3 December 1977), pp. 1001-1013.

12. Joel Darmstadter, Joy Dunkerley, and Jack Alterman, *How Industrial Societies Use Energy: A Comparative Analysis* (Baltimore: The Johns Hopkins University Press, 1977), pp. 185-207.

Chapter 9
Policy Alternatives

1. Lawrence J. White, "American Automotive Emissions Control Policy: A Review of the Reviews," *Journal of Environmental Economics and Management*, vol. 2, no. 4 (April 1976), pp. 231-246.

2. Richard O. Zerbe and Kevin Croke, *Urban Transportation for the Environment* (Cambridge, Mass.: Ballinger, 1975), pp. 66-67.

3. Frank P. Grad et al., *The Automobile and the Regulation of Its Impact on the Environment*. (Norman, Okla.: University of Oklahoma Press, 1975), pp. 273-276.

4. Thomas H. Maugh II, "Photochemical Smog: Is it Safe to Treat the Air?" *Science*, vol. 193 (3 September 1976), pp. 871-873.

5. William J. Baumol, "On Taxation and the Control of Externalities." *American Economic Review*, vol. 62 (June 1972), pp. 307-321.

6. Paul B. Downing and William D. Watson, Jr., "The Economics of Enforcing Air Pollution Controls," *Journal of Environmental Economics and Management*, vol. 1, no. 3 (November 1974), pp. 233-235.

7. See Robert M. Solow, "The Economist's Approach to Pollution and Its Control," *Science*, vol. 173 (6 August 1971), pp. 498-503. Also, Baumol, "Taxation and Control of Externalities," pp. 307-322. Also, U.S. Congress, Senate, Subcommittee on Air and Water Pollution of the Committee on Public Works, *Implementation of the Clean Air Act Amendments of 1970: Hearings, Part 3*, 92nd Congress, 2nd session, March 25, 1972, pp. 996-997, 1004-1005.

8. Richard Wilson, "Tax the Integrated Pollution Exposure," *Science*, vol. 178 (13 October 1972), pp. 82-83.

9. Thomas D. Crocker, "The Structuring of Atmospheric Pollution Control Systems," in Harold Wolozin, editor, *The Economics of Air Pollution* (New York: W.W. Norton & Company, Inc., 1966), p. 79.

10. J.H. Dales, *Pollution, Property & Prices* (Toronto: University of Toronto Press, 1968), pp. 92-97

11. Donald N. Dewes, *Economics and Public Policy: The Automobile Pollution Case* (Cambridge, Mass.: The MIT Press, 1974), pp. 94-95.

12. Zerbe and Croke, *Urban Transportation for the Environment*, p. 69.

13. Lead fouls up the catalytic emission control device used to meet the 1975-76 standards for mobile source emissions. See *Environmental Quality*, Third Annual Report of the Council on Environmental Quality (Washington: U.S. Government Printing Office, 1972), p. 112.

14. *Environmental Quality*, 1972, pp. 115-116, and H.R. 14931, 92nd Congress, May 11, 1972.

15. U.S. Department of Health, Education and Welfare, *Economic Impact of Air Pollution Controls on Gray Iron Foundry Industry* (Raleigh, N.C.: Environmental Health Service, 1970), p. 27.

16. John R. Meyer, "Transportation Solutions to the Energy 'Crisis,'" *National Bureau Report Supplement,* National Bureau of Economic Research, Inc., New York, February 1975, p. 3.

17. Meyer, "Transportation Solutions."

Index

About the Author

Charles T. Stewart, Jr. has been professor of economics at The George Washington University (from which he received the Ph.D. in 1954) since 1965. Before returning to The George Washington University as research professor in 1963, he taught economics at Utah State University and worked as a senior research analyst at The Georgetown University Graduate School, and as research economist and director of economic research at the United States Chamber of Commerce. He has written books on low-wage workers, minimum wages (with John H. Peterson), and the supply of medical personnel (with Corazon M. Siddayao), as well as numerous articles on manpower, regional and urban economics, and development. He has been a consultant to a number of federal government agencies.